Molecular Mechanisms and Genetics of Plant Resistance to Abiotic Stress

Molecular Mechanisms and Genetics of Plant Resistance to Abiotic Stress

Special Issue Editors

Jill M. Farrant
Maria-Cecília D. Costa

MDPI • Basel • Beijing • Wuhan • Barcelona • Belgrade

Special Issue Editors
Jill M. Farrant
University of Cape Town
South Africa

Maria-Cecília D. Costa
University of Cape Town
South Africa

Editorial Office
MDPI
St. Alban-Anlage 66
4052 Basel, Switzerland

This is a reprint of articles from the Special Issue published online in the open access journal *Plants* (ISSN 2223-7747) from 2018 to 2019 (available at: https://www.mdpi.com/journal/plants/special_issues/abiotic_stress).

For citation purposes, cite each article independently as indicated on the article page online and as indicated below:

LastName, A.A.; LastName, B.B.; LastName, C.C. Article Title. *Journal Name* **Year**, *Article Number*, Page Range.

ISBN 978-3-03928-122-0 (Pbk)
ISBN 978-3-03928-123-7 (PDF)

© 2020 by the authors. Articles in this book are Open Access and distributed under the Creative Commons Attribution (CC BY) license, which allows users to download, copy and build upon published articles, as long as the author and publisher are properly credited, which ensures maximum dissemination and a wider impact of our publications.

The book as a whole is distributed by MDPI under the terms and conditions of the Creative Commons license CC BY-NC-ND.

Contents

About the Special Issue Editors vii

Maria-Cecilia D. Costa and Jill M. Farrant
Plant Resistance to Abiotic Stresses
Reprinted from: *Plants* **2019**, *8*, 553, doi:10.3390/plants8120553 1

Leonardo Hinojosa, Juan A. González, Felipe H. Barrios-Masias, Francisco Fuentes and Kevin M. Murphy
Quinoa Abiotic Stress Responses: A Review
Reprinted from: *Plants* **2018**, *7*, 106, doi:10.3390/plants7040106 5

Llewelyn van der Pas and Robert A. Ingle
Towards an Understanding of the Molecular Basis of Nickel Hyperaccumulation in Plants
Reprinted from: *Plants* **2019**, *8*, 11, doi:10.3390/plants8010011 37

Ibrahim S. Elbasyoni, Sabah M. Morsy, Raghuprakash K. Ramamurthy and Atef M. Nassar
Identification of Genomic Regions Contributing to Protein Accumulation in Wheat under Well-Watered and Water Deficit Growth Conditions
Reprinted from: *Plants* **2018**, *7*, 56, doi:10.3390/plants7030056 49

Massimo Malerba and Raffaella Cerana
Effect of Selenium on the Responses Induced by Heat Stress in Plant Cell Cultures
Reprinted from: *Plants* **2018**, *7*, 64, doi:10.3390/plants7030064 64

Jun Wang, Ke Song, Lijuan Sun, Qin Qin, Yafei Sun, Jianjun Pan and Yong Xue
Morphological and Transcriptome Analysis of Wheat Seedlings Response to Low Nitrogen Stress
Reprinted from: *Plants* **2019**, *8*, 98, doi:10.3390/plants8040098 74

Simone Landi, Giorgia Capasso, Fatma Ezzahra Ben Azaiez, Salma Jallouli, Sawsen Ayadi, Youssef Trifa and Sergio Esposito
Different Roles of Heat Shock Proteins (70 kDa) During Abiotic Stresses in Barley (*Hordeum vulgare*) Genotypes
Reprinted from: *Plants* **2019**, *8*, 248, doi:10.3390/plants8080248 90

Pimprapai Butsayawarapat, Piyada Juntawong, Ornusa Khamsuk and Prakit Somta
Comparative Transcriptome Analysis of Waterlogging-Sensitive and Tolerant Zombi Pea (*Vigna vexillata*) Reveals Energy Conservation and Root Plasticity Controlling Waterlogging Tolerance
Reprinted from: *Plants* **2019**, *8*, 264, doi:10.3390/plants8080264 109

Erik Chovancek, Marek Zivcak, Lenka Botyanszka, Pavol Hauptvogel, Xinghong Yang, Svetlana Misheva, Sajad Hussain and Marian Brestic
Transient Heat Waves May Affect the Photosynthetic Capacity of Susceptible Wheat Genotypes Due to Insufficient Photosystem I Photoprotection
Reprinted from: *Plants* **2019**, *8*, 282, doi:10.3390/plants8080282 130

About the Special Issue Editors

Jill M. Farrant is a Full Professor and holds a South African Research Chair in "Systems Biology Studies on Plant Desiccation Tolerance for Food Security" in the Department of Molecular and Cell Biology, University of Cape Town. She is an acknowledged world leader in the field of plant desiccation tolerance (holding a rarely given A-rated status by the South African National Research Foundation), working on both seeds and resurrection plants. Her fundamental research involves the use of a systems biology approach (using techniques in molecular biology, biochemistry, cell biology, and physiology) to understand the protection mechanisms (and regulation thereof) laid down in orthodox seeds and vegetative tissue of resurrection plants, with the ultimate aim of using key protectants identified to make extremely drought-tolerant crops

Maria-Cecília D. Costa currently holds a Humboldt Research Fellowship for postdoctoral research at the Technical University of Munich (Germany). Her research focuses on the influence of natural variation in plant responses to abscisic acid (ABA). In 2017 and 2018, she was a postdoctoral fellow at the laboratory of Prof. Dr. Jill M. Farrant.

Editorial
Plant Resistance to Abiotic Stresses

Maria-Cecilia D. Costa [1,2,*] and Jill M. Farrant [2]

[1] Department of Plant Sciences, Technical University of Munich, 85354 Freising, Germany
[2] Department of Molecular and Cell Biology, University of Cape Town, Cape Town 7700, South Africa; jill.farrant@uct.ac.ya
* Correspondence: maceciliadc@gmail.com

Received: 15 November 2019; Accepted: 25 November 2019; Published: 28 November 2019

Abstract: Extreme weather events are one of the biggest dangers posed by climate breakdown. As the temperatures increase, droughts and desertification will render whole regions inhospitable to agriculture. At the same time, other regions might suffer significant crop losses due to floods. Usually, regional food shortages can be covered by surpluses from elsewhere on the planet. However, the climate breakdown could trigger sustained food supply disruptions globally. Therefore, it is necessary to develop more stress-resilient crop alternatives by both breeding new varieties and promoting underutilized crop species (orphan crops). The articles in this special issue cover responses of staple crops and orphan crops to abiotic stresses relevant under the climate breakdown, such as heat, water, high salinity, nitrogen, and heavy metal stresses. This information will certainly complement a toolkit that can help inform, support, and influence the design of measures to deal with the climate crisis.

Keywords: water stress; high salinity stress; heat stress; orphan crop; nickel hyper-accumulation

---

There is a general consensus that global heating is occurring, with predictions that average surface temperatures will increase by roughly 0.2 °C per decade over the next 30 years. This, in turn, is predicted to result in increased frequencies of extreme weather events, drastically reducing crop yields [1]. This is already in evidence in the altered start and end of growing seasons over the past 20 years, which has contributed to regional crop yield reductions, reduced freshwater availability, and has exacerbated the loss of biodiversity [2]. These effects are not uniformly distributed among regions, countries, and land-management systems. It is proposed that European Mediterranean countries are likely to experience more frequent droughts and warmer summers, and temperature increases in North America will result in a shift in the start of the rainy season from spring to winter [3]. El Niño events might grow more frequent, bringing more rain to South American coastal areas and less rain to Indonesia, the Pacific Rim, and Australia [3]. The severity of monsoon rains in Asia might increase, causing more floods in the lowlands [3]. Warmer temperatures will speed up the melting of permafrost in Siberia and increase desertification in Northern Africa [3]. These factors, in combination with elevated salinity in groundwater, poor agricultural practices, and human-related disturbances, are resulting in a considerable reduction of agricultural yield and, thus, local (and inevitably global) food insecurity. This Special Edition reports on evidence of the effects of such abiotic stressors on select crops, and we summarize potential solutions to mitigate food security as a consequence of abiotic stressors.

As one of the primary aspects of climate change, heating itself has a direct bearing on plant mortality. Studies show that global cereal production was reduced by 6.2% between 2000 and 2007 due to the effects of heat and, consequently, drought [4]. In this special issue, Chovanccek et al. (2019) show that heat waves decrease photosynthetic activity and stomatal conductance in wheat. Interestingly, after control conditions were reestablished, plants showed persistent reductions in photosynthesis, suggesting the action of a protective mechanism to prevent the collapse of photoprotective functions [5].

Malerba and Cerana (2019) show that selenium reduces cell death associated with heat stress in tobacco cell cultures by reducing the accumulation of reactive oxygen species, peroxidation of membrane lipids, programmed cell death, and accumulation of stress-related proteins (i.e., BiP, 14-3-3, cytochrome c) [6].

Drought has always been the major cause of crop loss, and as discussed above, this is being exacerbated by global heating. In wheat, drought stress leads to increased grain protein content, but with a yield penalty [7]. In this issue, Elbasyoni et al. (2018) utilizes wheat genetic resources to identify loci correlated with grain protein content under drought stress and well-watered conditions. They identify loci that can aid the development of marker-assisted selection for grain protein content, minimizing the yield penalty and leading to a better understanding of the genetic architecture that controls these traits in wheat [7].

Barley is a staple cereal grown in temperate climates globally. High-yielding cultivars generated through breeding have lost tolerance to drought, salt, and heavy metal stress present in the wild ancestor *Hordeum spontaneum* [8]. Stress-tolerant barley landraces from the semi-arid Mediterranean region represent an important source of genetic diversity for the improvement of high-yield cultivars. In this special issue, Landi et al. (2019) investigates how different barley landraces respond to salinity and osmotic stress. Their results can be useful for breeding strategies to reintroduce stress tolerance in cultivated varieties [8].

Recently, there has been increased interest in the use of orphan crops, considered to be minor on a global scale, yet important locally for food security in the developing world [9]. Most such crops have greater inherent stress tolerance or resistance than current staple cereals (wheat, rice, and maize) in which such characteristics have been lost by conventional breeding for seed size and yield under well-watered conditions [10]. They are, thus, good candidates to include in molecular breeding programs. However, in many instances, the lack of genomic resources on such species has hindered progress in this regard. Quinoa is an Andean orphan crop adapted to a wide range of marginal agricultural areas, including those prone to drought and high soil salinity. In this special issue, Hinojosa et al. (2018) review the resources available on this species in relation to its tolerance to abiotic stressors [11].

Excessive rainfall as a consequence of global heating is also an increasing threat to food security. It has been reported to be the second-largest contributor to loss of maize yields in the United States, totaling damage of 10 billion USD from 1989 to 2016 [12]. Flooding creates low oxygen environments at the root level and causes an imbalance between the slow diffusion of gases in water compared to air and high consumption by microorganisms and roots, leading to oxygen-starved plants [13]. In this issue, Butsayawarapat et al. (2019) characterize mechanisms controlling water-logging tolerance in the orphan crop *Vigna vexillata* (also known as zombi pea or tuber cowpea) [13]. Zombi pea is closely related to cowpea and is cultivated for their fleshy tubers and mature seeds. One of the mechanisms identified is the increased expression of aquaporins, proteins that represent an important selective pathway for water to move across cellular membranes.

Because weather extremes are clearly impacting crop yield, it is also important to take alternative measures to meet the target of maintaining global average temperatures below 2 °C above pre-industrial levels. This requires several actions already in place and additional non-climate actions that can deliver a climate benefit, such as a significant reduction of nitrogen (N) and heavy metal pollution [14]. N pollution has become a major contributor to environmental stresses this century, its main source being an inefficient use and management of synthetic fertilizers and manure by the agricultural sector [14]. Therefore, it is urgent to understand plant responses to N fertilization and seek ways to improve plant N uptake. In this special issue, Wang et al. (2019) investigate the molecular responses of wheat seedlings grown under low N conditions. Their results can be useful to understand the lower limits of N input tolerated by staple crops and for the design of innovative approaches to agricultural N management [15].

Heavy metal pollution might be aggravated by a climate breakdown due to the release of these metals from sediments to water reservoirs [16]. Remediation of heavy metals by conventional methods is challenging and generates several secondary wastes. On the other hand, bio-remediation using plants and microorganisms represents a more promising solution. Bio-remediation involves the adsorption, reduction, or removal of contaminants from the environment by hyper-accumulator plants and microorganisms [16]. In this special issue, van der Pas and Ingle (2019) review the molecular basis of nickel hyper-accumulation in plants and suggest potential future avenues of research in this field [17]. Although nickel is essential for plant growth, extremely high soil concentrations have left farmlands unsuitable for agriculture. Therefore, a better understanding of its functional roles, as well as hyper-accumulation mechanisms, is essential for the formulation of bio-remediation strategies.

The climate crisis is already causing yield losses of major crops and the reduction of arable land worldwide. Hence, to meet our goals of food security, we need to increase the production of food crops that can withstand the ongoing and future environmental changes. This must be done by both developing new varieties of current staple crops and the use of alternative (orphan) crops that produce a harvestable yield in arid and semi-arid regions. At the same time, we must mobilize a wide range of public and private actors to implement measures that reduce emissions, strengthen initiatives already underway, and mitigate the climate crisis. We hope that the articles in this special issue will complement the set of tools that can help inform, support, and influence the design of such measures.

Conflicts of Interest: The authors declare no conflict of interest.

References

1. Bailey-Serres, J.; Parker, J.E.; Ainsworth, E.A.; Oldroyd, G.E.D.; Schroeder, J.I. Genetic strategies for improving crop yields. *Nature* **2019**, *575*, 109–118. [CrossRef] [PubMed]
2. Brown, C.; Alexander, P.; Arneth, A.; Holman, I.; Rounsevell, M. Achievement of Paris climate goals unlikely due to time lags in the land system. *Nat. Clim. Chang.* **2019**, *9*, 203–208. [CrossRef]
3. Hopkin, M. Climate change: World round-up. *Nature* **2005**. [CrossRef]
4. Lesk, C.; Rowhani, P.; Ramankutty, N. Influence of extreme weather disasters on global crop production. *Nature* **2016**, *529*, 84–87. [CrossRef] [PubMed]
5. Chovancek, E.; Zivcak, M.; Botyanszka, L.; Hauptvogel, P.; Yang, X.; Misheva, S.; Hussain, S.; Brestic, M. Transient Heat Waves May Affect the Photosynthetic Capacity of Susceptible Wheat Genotypes Due to Insufficient Photosystem I Photoprotection. *Plants* **2019**, *8*, 282. [CrossRef] [PubMed]
6. Malerba, M.; Cerana, R. Effect of selenium on the responses induced by heat stress in plant cell cultures. *Plants* **2018**, *7*, 64. [CrossRef] [PubMed]
7. Elbasyoni, I.S.; Morsy, S.M.; Ramamurthy, R.K.; Nassar, A.M. Identification of genomic regions contributing to protein accumulation in wheat under well-watered and water deficit growth conditions. *Plants* **2018**, *7*, 56. [CrossRef] [PubMed]
8. Landi, S.; Capasso, G.; Ben Azaiez, F.E.; Jallouli, S.; Ayadi, S.; Trifa, Y.; Esposito, S. Different Roles of Heat Shock Proteins (70 kDa) During Abiotic Stresses in Barley (Hordeum vulgare) Genotypes. *Plants* **2019**, *8*, 248. [CrossRef] [PubMed]
9. Tadele, Z. Orphan crops: their importance and the urgency of improvement. *Planta* **2019**, *250*, 677–694. [CrossRef] [PubMed]
10. Landi, S.; Hausman, J.-F.; Guerriero, G.; Esposito, S. Poaceae vs. abiotic stress: Focus on drought and salt stress, recent insights and perspectives. *Front. Plant Sci.* **2017**, *8*, 1214. [CrossRef] [PubMed]
11. Hinojosa, L.; González, J.A.; Barrios-Masias, F.H.; Fuentes, F.; Murphy, K.M. Quinoa abiotic stress responses: A review. *Plants* **2018**, *7*, 106. [CrossRef] [PubMed]
12. Li, Y.; Guan, K.; Schnitkey, G.D.; DeLucia, E.; Peng, B. Excessive rainfall leads to maize yield loss of a comparable magnitude to extreme drought in the United States. *Glob. Chang. Biol.* **2019**, *25*, 2325–2337. [CrossRef] [PubMed]
13. Butsayawarapat, P.; Juntawong, P.; Khamsuk, O.; Somta, P. Comparative Transcriptome Analysis of Waterlogging-Sensitive and Tolerant Zombi Pea (Vigna Vexillata) Reveals Energy Conservation and Root Plasticity Controlling Waterlogging Tolerance. *Plants* **2019**, *8*, 264. [CrossRef] [PubMed]
14. Kanter, D.R. Nitrogen pollution: A key building block for addressing climate change. *Clim. Change* **2018**, *147*, 11–21. [CrossRef]

15. Wang, J.; Song, K.; Sun, L.; Qin, Q.; Sun, Y.; Pan, J.; Xue, Y. Morphological and transcriptome analysis of wheat seedlings response to low nitrogen stress. *Plants* **2019**, *8*, 98. [CrossRef] [PubMed]
16. Jacob, J.M.; Karthik, C.; Saratale, R.G.; Kumar, S.S.; Prabakar, D.; Kadirvelu, K.; Pugazhendhi, A. Biological approaches to tackle heavy metal pollution: A survey of literature. *J. Environ. Manag.* **2018**, *217*, 56–70. [CrossRef] [PubMed]
17. Van der Pas, L.; Ingle, R.A. Towards an understanding of the molecular basis of nickel hyperaccumulation in plants. *Plants* **2019**, *8*, 11. [CrossRef] [PubMed]

© 2019 by the authors. Licensee MDPI, Basel, Switzerland. This article is an open access article distributed under the terms and conditions of the Creative Commons Attribution (CC BY) license (http://creativecommons.org/licenses/by/4.0/).

Review

Quinoa Abiotic Stress Responses: A Review

Leonardo Hinojosa [1,2], Juan A. González [3], Felipe H. Barrios-Masias [4], Francisco Fuentes [5] and Kevin M. Murphy [1,*]

1. Sustainable Seed Systems Lab, Department of Crop and Soil Sciences, College of Agricultural, Human, and Natural Resource Sciences, Washington State University, Pullman, WA 99164-6420, USA; l.hinojosasanchez@wsu.edu
2. Facultad de Recursos Naturales, Escuela de Agronomía, Escuela Superior Politécnica de Chimborazo, Riobamba 060106, Ecuador
3. Fundación Miguel Lillo, Instituto de Ecología, Miguel Lillo, San Miguel de Tucumán Post 4000, Argentina; jalules54@gmail.com
4. Department of Agriculture, Veterinary and Rangeland Sciences, University of Nevada-Reno, Reno, NV 89557, USA; fbarrios@cabnr.unr.edu
5. Facultad de Agronomía e Ingeniería Forestal, Pontificia Universidad Católica de Chile, Vicuña Mackenna, Macul, Santiago 4860, Chile; frfuentesc@uc.cl
* Correspondence: kmurphy2@wsu.edu; Tel.: +1-509-335-9692

Received: 4 November 2018; Accepted: 26 November 2018; Published: 29 November 2018

Abstract: Quinoa (*Chenopodium quinoa* Willd.) is a genetically diverse Andean crop that has earned special attention worldwide due to its nutritional and health benefits and its ability to adapt to contrasting environments, including nutrient-poor and saline soils and drought stressed marginal agroecosystems. Drought and salinity are the abiotic stresses most studied in quinoa; however, studies of other important stress factors, such as heat, cold, heavy metals, and UV-B light irradiance, are severely limited. In the last few decades, the incidence of abiotic stress has been accentuated by the increase in unpredictable weather patterns. Furthermore, stresses habitually occur as combinations of two or more. The goals of this review are to: (1) provide an in-depth description of the existing knowledge of quinoa's tolerance to different abiotic stressors; (2) summarize quinoa's physiological responses to these stressors; and (3) describe novel advances in molecular tools that can aid our understanding of the mechanisms underlying quinoa's abiotic stress tolerance.

Keywords: quinoa; abiotic stress; heat; drought; salinity; mechanism

---

1. Introduction

Abiotic stress is the primary cause of crop losses, decreasing yields by more than 50% worldwide [1]. Many of these stressors naturally occur in combination. The main abiotic stresses—drought, waterlogging, high salinity, heavy metals, excess heat, frost, and ultraviolet-B light irradiance (UV-B)—have been extensively studied in plants [1–8]. The average annual global air temperature is expected to increase between 0.3 and 0.7 °C per decade, and by the end of this century, the highest predicted temperature increase approximates 4.8 °C due to climate change [9]. In this scenario, the predicted temperature extremes, or heat waves in summer, have received more attention due to their anticipated adverse impacts on human mortality, economies, and ecosystems [10,11].

Quinoa is an Andean crop, known as "kiuna" or "kinwa" in the Quechua language and "jupha" or "jiura" in the Aymara language [12]. Quinoa is widely cultivated, from sea level at the coast to 4000 m above sea level (m.a.s.l.). The plant's natural geographical distribution ranges from southern Colombia (2 °N) to the coast of south-central Chile (43 °S), including a branch in northwest Argentina and some subtropical lowlands in Bolivia [13,14] (Figure 1). Originally, quinoa was domesticated in southern Peru and Bolivia close to Titicaca Lake and evidence of human cultivation dates back to between 8000 and 7500 years before present (B.P.) [15].

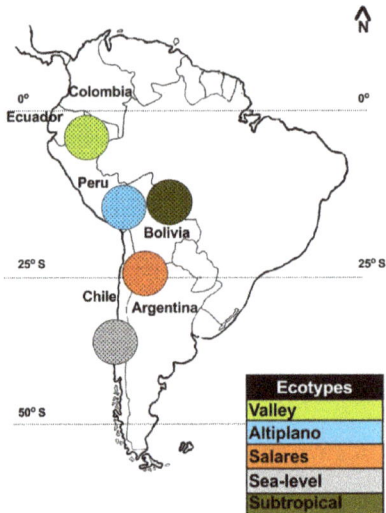

Figure 1. Geographic distribution of the five quinoa ecotypes.

Quinoa is traditionally classified into five ecotypes, based on geographic adaptation, as follows: (1) valley = grown at 2000 to 3500 m.a.s.l. in Colombia, Ecuador, Peru, and Bolivia; (2) altiplano = grown at high altitudes of more than 3500 m.a.s.l. around Titicaca Lake on the border of Bolivia and Peru; (3) salares = grown in the salt flats of Bolivia and Chile and has a high tolerance to salinity; (4) sea-level = grown in the low-altitude areas of southern and central of Chile; and (5) subtropical or yungas = grown in the low-altitude, humid valleys of Bolivia and includes late-flowering genotypes (Figure 1) [12]. Quinoa germplasm is highly diverse. The natural variability in different traits, such as inflorescence type, seed color, seed size, life-cycle duration, salinity tolerance, saponin content, and nutritional value, allows quinoa to adapt to diverse environments [16–23]. To protect the genetic variability of quinoa in the Andean region, several gene banks have been created since the 1960s. A total of 16,422 accessions have been conserved in 59 gene banks across 30 countries, the majority of which are concentrated in Bolivia and Peru [23–25].

Quinoa is adapted to a wide range of marginal agricultural soils, including those with high salinity and those prone to drought. Recently, several papers have primarily addressed salt and drought tolerance in quinoa [14,26–33]. However, since the quinoa reference genome has been published [34], new transcriptome studies in salinity and drought in quinoa have been completed. Furthermore, information is limited about quinoa's tolerance to other abiotic stress factors, such as frost, UV-B irradiance, and high air temperature. The purpose of this review is to give an overview of the current state of knowledge about quinoa tolerance to salt and drought, plus a variety of other abiotic stressors, namely high air temperature, UV-B radiation, frost, waterlogging, and heavy metal contamination. In particular, we discuss: (1) quinoa's morphological, physiological, and molecular responses to these stressors; (2) management strategies to reduce the effects of these stressors; and (3) recent advancements in genetic and molecular resources that can help breeders improve quinoa's tolerance to abiotic stress.

2. Drought

Agricultural drought is defined as the insufficient soil moisture that causes a reduction in plant production [35]. Quinoa is considered a drought-tolerant crop, capable of growing and producing seed grain in the semi-desert conditions of Chile, the arid mountain regions of northwest Argentina, and the Altiplano area of Peru and Bolivia. These environments are characterized as extremely arid, with less than 200 mm of annual rainfall [19,36–41]. Quinoa can also adapt and produce seed in semi-arid and arid environments outside of the Andean region, such as Asia, North Africa, the Near East, and the Mediterranean [28,42–45].

Although quinoa is inherently drought tolerant, different climatic models predict an increase in drought frequency, especially in the altiplano region of the Andes, where quinoa is grown traditionally by small farmers [46]. Thus, understanding the drought response mechanisms in quinoa is critical for developing varieties with improved drought tolerance.

2.1. Drought Response Mechanisms in Quinoa

Plants develop different response mechanisms to endure a lack of water. These mechanisms can be divided into three groups: (1) morphological strategies, such as avoidance, for instance, deeper roots, and phenotypic flexibility related to ontogenic processes that can contribute to the scape and avoidance strategy; (2) physiological strategies, such as antioxidant defense, cell membrane stabilization, plant growth regulation, stomatal conductance, and osmotic adjustment; and (3) molecular strategies, such as activating stress proteins (osmoprotectants) and aquaporins [47].

Quinoa's flowering and milk grain stages have been established as the most drought-sensitive [38]. Several studies have been conducted to understand the quinoa plant's mechanisms under drought stress [39,48–54]. In a pot experiment under drought conditions, Jacobsen et al. (2009) reported an increased concentration of abscisic acid (ABA) in the roots of quinoa altiplano variety 'INIA-Illpa', which induced a decreased turgor of stomata guard cells [52]. The same mechanism was observed in the leaves of sea-level variety 'Titicaca' when plants were grown under water deficient and control conditions [51,54]. Furthermore, during drought stress of 'Titicaca', the concentration of xylem ABA increased faster in the shoots than the roots [55]. Similar results were observed again in 'Titicaca' and altiplano variety 'Achachino'; xylem sap ABA concentration increased two days after drought treatment and decreased to the control levels after re-watering. Under the drought conditions, 'Titicaca' had higher ABA concentrations than 'Achachino' [56].

Other drought response mechanisms in quinoa are the synthesis of reactive oxygen species (ROS) scavengers; accumulation of osmolytes as an antioxidant defense, particularly ornithine and raffinose pathways; and the accumulation of soluble sugars and proline, which adjust cellular osmotic potential [49,57–59]. Quinoa also develops response mechanisms to reduce water loss through rapid stomatal closure, cellular water deficit regulation, and root-to-shoot ratios that trigger a high water-use efficiency [50–52,54–56,60,61]. However, Jensen et al. (2000) found that quinoa altiplano variety 'Kankolla' was insensitive to drought relative to stomatal response at early growth stages [53]. In response to this finding, they proposed that high net photosynthetic rates and a specific leaf area in early growth stages support water uptake by larger root systems that helps the plant avoid drought later on. Other drought response mechanisms could involve a delay in development when drought was imposed at the pre-anthesis stage under Bolivian Altiplano conditions [62].

One of the major effects of drought on plants is a reduction in the photosynthetic rate, which is primarily due to stomatal closure [63]. Leaf gas exchange and carbon isotope discrimination (Δ) are common approaches used to study plants under drought conditions [64]. González et al. (2011) evaluated leaf gas exchange and $\Delta^{13}C$ in 10 quinoa genotypes grown in the arid mountain region of northwest Argentina, which receives 160 mm of rainfall during the growing season [39]. Results showed that quinoa genotypes with higher stomatal conductance were capable of maintaining higher photosynthetic rates. Additionally, the researchers observed high variability in the grain yield among genotypes and found a positive correlation between $\Delta^{13}C$ and yield. Similar results were reported by [60], where drought-induced quinoa experienced pronounced stomatal and mesophyll limitations to CO_2 transport. However, in quinoa greenhouse experiments, indicators of leaf photosynthetic capacity, such as the maximum quantum yield of PSII (F_v/F_m) and quenching analysis (qP and qN), were insensitive to water stress [50,60,65]. On the other hand, F_v/F_m decreased in response to drought effects in the variety 'Titicaca' in a greenhouse experiment [54].

During two successive quinoa growing seasons in Morocco, field experiments used OJIP analysis, defined by the O, J, I, and P steps that correspond to the redox states of the photosystem (PS II and PS I); this analysis explores changes in the photosystem II (PSII) photochemical performance [66]. Results showed that drought stress in a sea-level grown 'Puno' variety induced a decrease in F_v/F_m and in the quantum yield of electron transport ($\varphi E0$), proposing OJIP parameters as a viable drought stress evaluation tool. Conversely, another analysis of chlorophyll a fluorescence OJIP transient in the sea-level variety 'Red Head', grown in semi-controlled conditions in Italy, revealed no difference in 16 chlorophyll fluorescence parameters between control and drought treatments [60]. These two contradictory studies likely demonstrate genotypic variability in the quinoa response to drought. Consequently, additional studies that include a greater number of genotypes and simultaneous measurements of gas exchange are required to establish chlorophyll a fluorescence OJIP transient's effectiveness as a drought evaluation tool for quinoa.

Finally, root system architecture and its relationship to soil moisture conditions has been studied in quinoa and quinoa relatives. Quinoa roots exhibit faster elongation and abundant and longer external branching of the roots that improve their foraging capacity compared to quinoa relatives *C. hiricinum* and *C. pallidicaule* [67,68]. Recently, a root system architecture and dynamics study was conducted in drought conditions, comparing *C. hiricinum* and *C. pallidicaule*, a rainy-habitat and a dry-habitat quinoa genotype, respectively. Results showed that the quinoa genotypes exhibited accelerated taproot growth in dry soil conditions compared to the other two species. Furthermore, the quinoa genotype from the dry habitat showed longer, coarser, and more numerous root segments than the wet-habitat genotype [48]. These findings led the authors to suggest quinoa as a promising plant model to investigate biophysical and ecophysiological traits of plant rooting in deep soil layers.

2.2. Field Studies Under Drought Conditions

In Italy, Pulvento et al. (2012) found that yields for 'Titicaca' ranged between 2.30–2.70 t ha^{-1} whether grown under high irrigation (300–360 mm) or deficit irrigation (200–220 mm) during the growing season [44]. Thus, the study concluded no significant yield reductions due to a lack of water. Similar results in the yield with the same variety were observed in Demark when plants were grown under irrigated and non-irrigated conditions in sand, sandy loam, and sandy clay loam soils [69]. However, in Egypt, five quinoa genotypes were evaluated under three different growing season water regimes, consisting of high (820 mm), moderate (500 mm), and low (236 mm) irrigation treatments (rainfall + irrigation). Results showed high variability in morphological traits and yield among genotypes across the different water regimes. For instance, sea-level variety 'QL-3' exhibited the biggest reduction in yield (56%) under severe water stress; valley variety 'CICA-17' showed the smallest reduction (12%) [70].

Quinoa yield reductions under dryland conditions were reported for sea-level varieties 'Cherry Vanilla' and 'Oro de Valle' when grown in an organic field in Pullman, WA—an area characterized by dry, warm summers [71]. This study was carried out under three intercrop treatments (clover/medic mix, fescue grass/clover mix, and a no intercrop control) and three irrigation regimens (dryland, 64 mm, and 128 mm of water). Results showed that neither of the two intercrop treatments affected the quinoa yield under irrigated or non-irrigated conditions. However, irrigation can relieve the effect of high temperature; for instance, the mean yield under the dryland regime increased from 0.2 t ha^{-1} to 1.2 t ha^{-1} with an extra 128 mm of water during the growing season [71].

2.3. Irrigation Strategies to Mitigate Drought Stress

The crop coefficient (K_c) is the ratio of evapotranspiration of a crop to a reference crop, such as a perennial grass. K_c helps predict crop irrigation needs in different phenological development stages using meteorological data from a weather station [72]. To estimate irrigation requirements for the quinoa altiplano variety 'Chucapaca' grown in the Bolivian Altiplano, Garcia et al. (2003) calculated K_c as 0.52, 1.00, and 0.70 for the initial, mid-season, and late-season phenological stages, respectively [73]. In contrast, K_c values for the variety 'Titicaca' grown in Demark were higher than values reported for the Bolivian Altiplano, equaling 1.05, 1.22, and 1.00 for initial, mid-season, and late-season stages, respectively [69]. Full irrigation to increase the quinoa yield in water-scarce regions is not an option; however, partial root zone drying and deficit irrigation are practices that reduce the amount of water used during the growing season without detriments to yield and might be useful alternatives [38,74]. Nevertheless, in regions with poor quality water (saline groundwater table), deficit irrigation must be managed with caution to avoid high salt accumulation in the root zone [75].

2.4. Other Drought Stress Mitigation Strategies

In addition to irrigation, several other approaches for relieving drought stress in quinoa have been studied. For example, greenhouse experiments with 'Titicaca' showed that applications of nitrogen (N) supplied as ammonium nitrate (NH_4NO_3) at a rate of 0.6 g N plot^{-1} could improve plant performance under water stress. Observed drought tolerance mechanisms included faster stomatal closure, lower leaf water potential, and higher leaf ABA concentrations [76]. Other studies of drought stressed quinoa have found that adding compost and acidified biochar to soils under drought conditions can improve quinoa plant growth, yield, physiological, and antioxidant activity, and chemical and biochemical attributes of quinoa seeds [77–80]. For example, under field conditions in Morocco, organic amendments can relive the drought effect in quinoa; yields increased from 1.7 to 2.0 t ha^{-1} using 10 t ha^{-1} of compost under non-irrigated conditions [78,79]. Similar increases in yield were observed for two quinoa genotypes grown in the semi-arid conditions of Chile when vermicompost was used to enhance soil organic matter [40]. The yield increased from 5.8 g plot^{-1} to 9.4 g plot^{-1} by adding acidified biochar in drought conditions [80,81].

Moreover, a study with quinoa genotype 'V$_9$', subjected to varying irrigation regimes, demonstrated that foliar applications of 150 mg L^{-1} synthetic ascorbic acid and 25% concentration of orange juice (natural ascorbic acid) diluted in distilled water mitigated the harmful effects of drought stress in quinoa [77]. Plant growth, total carotenoids, free amino acids, and several antioxidant enzymes increased due to synthetic ascorbic acid and orange juice in drought conditions [77]. Exogenous ascorbic acid protects lipids and proteins from the plants against drought-induced oxidative adversaries [82]. Proline was used as another foliar treatment under field conditions in Egypt [83]. Results showed that foliar applications of 12.5 mM and 25.0 mM of proline improved growth parameters, relative water content, yield components, and nutritional values. Applications of 25.0 mM of proline increased the yield from 6.23 g plant^{-1} to 8.56 g plant^{-1} in drought conditions. Additionally, a pot experiment under greenhouse conditions using the quinoa sea-level variety 'Pichaman' showed that applying 80 mM of exogenous H_2O_2 as a seed primer and 15 mM as a foliar spray improved the quinoa performance under drought conditions. For instance, plants exhibited higher photosynthetic rates, stomatal conductance, chlorophyll content indices, proline levels, sugar contents, and ABA regulation [84]. Exogenous H_2O_2 acts as an oxidative modifier and mobilizer of stored proteins [85].

Unique fungal-root associations in quinoa may aid in the plant's ability to tolerate drought conditions. Several studies have characterized the endophytic fungi associated with quinoa roots and bacterial endophytes in quinoa seeds [86–89]. Quinoa roots were collected in natural conditions close to the Salt Lake of the Atacama Desert in Chile. Molecular analysis showed that quinoa roots shelter a diverse group of endophytic fungi. *Penicillium*, *Phoma*, and *Fusarium* genera dominated the fugal community [90]. The fungus *Penicillium minioluteum*, isolated from the characterization described above, was used to study the effects of root endophytic fungi on drought stress in a quinoa variety from the Atacama Desert. Results demonstrated a 40% improvement in root biomass relative to the treatment with no inoculum. However, the study found no improvement in photosynthesis, stomatal conductance, or photochemical efficiency by the presence of the endophytic fungi. Thus, the interaction between *P. minioluteum* and quinoa exhibited a positive response in root biomass, but only under drought conditions [86]. Another study was conducted using the root endophyte *Piriformospora indica* and the quinoa valley variety 'Hualhuas' under greenhouse conditions. Results showed the successful colonization of *P. indica* in quinoa. This association could mitigate some drought effects by improving the plant water and nutrient status, resulting in the capacity to increase total biomass, stomatal conductance, leaf water potential, and net photosynthesis [87].

2.5. Seed Quality Under Water Limitations

Environmental and climatic factors influence the nutritional quality of quinoa seeds. Variations in amino acids, protein content, mineral composition, and phytate were observed in 10 quinoa varieties between two semi-arid locations, northwest Argentina and the Bolivian Altiplano [91]. The interaction between genotype and environment (G × E) was responsible for mineral composition, amino acids, and protein variations among quinoa varieties [91,92]. Similar results were found with three quinoa varieties planted in Chile, Argentina, and Spain. For example, seed quality was primarily dependent upon G × E, with the exception of saponin and fiber content, which were more stable across locations [93].

In another study with two sea-level varieties, 'Cherry Vanilla' and 'Oro de Valle', seed protein content increased when quinoa was grown with irrigation and a clover-medic mixture intercrop system, compared to the same intercrop system without irrigation. Furthermore, the irrigated plants exhibited increased seed concentrations of P, Mg, and Fe, but decreased concentrations of Ca, Cu, and Zn, compared to the non-irrigated treatment [71]. On the other hand, Pulvento et al., (2012) found no differences in any aspect of seed quality in 'Titicaca' among three different irrigation treatments. However, seed fiber and saponin content increased when the quinoa plants were well-irrigated, compared to plants without irrigation [44].

In the south-central zone of Chile, seed quality was evaluated in quinoa sea-level varieties 'Regalona', 'AG2010', and 'B080', grown under four regimes of water availability in both greenhouse and field experiments. Results showed an increase in the seed antioxidant capacity of all three varieties and a minimal reduction in seed yield in 'AG2010' under 20% soil water availability relative to 95% soil water availability [94]. Moreover, 20% soil water availability in 'AG2010' increased globulin content, and the effect of washing quinoa seeds with water changed the concentration and electrophoretic pattern of albumins and globulins [95].

2.6. Gene Expression Under Water Limitation

Raney et al. (2014) performed the first RNA sequencing (RNA-seq) transcriptome analysis on quinoa under drought conditions with two varieties: valley variety 'Ingapirca' and salares variety 'Ollague'. 'Ollague' demonstrated a greater drought tolerance compared to 'Ingapirca', based on several physiological parameters, including stomatal conductance, photosynthetic rate, and stem water potential [96]. RNA-seq using root tissue from both varieties under a control and different water stress treatments, identified 462 differentially expressed contigs and 27 putative genes with regulatory functions based in the interaction terms. Several of these 27 genes have unknown protein functions. However, other genes, such as AUR62041909 and AUR62015321, have known functions. Specifically, gene AUR62041909 functions as an intermediate in the biosynthesis of flavonoids in plants. Gene AUR62015321 belongs to the dirigent family of proteins, which are induced during the disease response in plants and are involved in lignification [97].

Heat-shock proteins (HSPs) have been studied since the 1960s as one of the stress-inducible proteins under several types of stress, but mostly under lethal temperatures [98]. In recent decades, HSPs have gained more attention as molecular chaperones, preventing the accumulation of other proteins and playing an important role in protein folding [99]. HSP superfamilies are grouped based on molecular weight, for example, HSP100, HSP90, HSP70, HSP60, and small heat-shock proteins (sHSPs) [98]. The HSP70s are not only up-regulated during heat stress conditions in plants, but also play an important role in response to other stresses, such as drought [100]. Recently, Liu et al., (2018) identified and characterized 16 quinoa HSP70 members (*Cqhsp70s*) in the newly sequenced quinoa genome [34], based on HSP70s in *Arabidopsis* [101]. Their study analyzed the expressions of 13 *Cqhsp70s* genes under drought conditions induced by polyethylene glycol 6000 (PEG6000). Results showed a significant variation in the gene response to drought stress. For instance, the expression of six out the 13 *Cqhsp70s* genes was down-regulated at the beginning of drought stress and during the recovery time. In another example, the expression of the gene *AUR62024018* remains high throughout the duration of the drought treatment. Moreover, one-half of the genes evaluated exhibited a "drop-climb-drop" expression pattern, which was similar to the homolog genes in *Arabidopsis*.

In other work, Morales et al., (2017) studied the transcriptional responses of quinoa under drought stress [102]. First, they found that the salares variety 'R49' presented the highest drought tolerance compared to 'PRJ' and BO78 sea-level varieties; 'R49' displayed the best performance on physiological parameters, such as relative water content, electrolyte leakage, and (F_v/F_m). Second, RNA-seq was carried out on 'R49' under control and drought conditions. Fifty-four million reads were obtained for the control and 51 million for drought conditions. All reads were assembled into 150,952 contigs; 19% of genes (306 contigs) were not represented in published databases of homologous genes. Fifteen target genes were selected to analyze gene expression. Some of these genes were selected based on other plant models in which these genes have been induced under drought stress, focusing on ABA biosynthesis and ABA transport pathway functions. Other target genes were selected based on those that exhibited changes of representation reads by RNA-seq in quinoa. Results showed that just two genes associated with ABA biosynthesis, CqNCED3a and CqNCDE3b, which are localized in plastids, were up-regulated in response to drought in quinoa. Moreover, all genes that exhibited changes from representation reads, *CqHSP20* (putative chaperones hsp20- protein superfamily), *CqCAP160* (cold acclimation protein 160), *CqLEA* (late embryogenesis abundant protein family protein), *CqAP2/ERF* (integrase-type DNA-binding protein superfamily), *CqPP2C* (protein phosphatase protein family 2c), *CqHSP83* (chaperone protein, protein family HTPG), and *CqP5CS* (delta 1-pyrroline-5-carboxylate synthase 2), were up-regulated. In particular, genes CqHSP20 and CqLEA were altered over 140-fold expression. Both HSP studies described above concur that HSPs play an important role in the adaptation of quinoa under drought stress. Thus, quinoa could be an excellent model species to study HSPs under multiple stresses, such as drought, heat, and salinity.

3. Salinity

Salinity refers to the presence of the major dissolved inorganic solutes, mainly Na^+, Mg^{2+}, Ca^{2+}, K^+, Cl^-, SO_4^{2-}, HCO_3^-, and CO_3^{2-}. Soil salinity refers to the soluble plus readily dissolvable salts in the soil or in the aqueous extract and is quantified as the total concentration of the salts or by measuring the electrical conductance (EC) of a saturation soil extract. Soils are considered saline when EC is more than 4 dS m^{-1} at 25 °C [103]. The response of quinoa to salinity has been studied intensely during the last 20 years, prior to May 2018. Over 120 studies on the relationship between salinity and quinoa were published between 1998 and 2018; of these, approximately 60% of the studies were published within the last five years. Three extensive reviews about quinoa as a model for understanding salt tolerance have been published [26,33,104]. In the following, we briefly summarize the response mechanisms associated with quinoa's salt tolerance and then detail new advances in gene transcription for understanding quinoa's response to high salinity.

High salinity is considered one of the major abiotic stresses that limit crop yields because it causes reduced photosynthesis, respiration, and protein synthesis. Photosynthesis reduction, membrane denaturalization, nutrient imbalance, stomatal closure, and a dramatic increase in ROS production are the main physiological changes in plants under salinity stress [105–107]. High accumulations of ROS cause serious plant toxicity, including oxidative damage in proteins, lipids, and DNA; whereas, low concentrations of ROS act as signaling molecules [108–111].

3.1. Tolerance and Effects of High Salinity in Quinoa

Quinoa has been identified as a facultative halophyte crop, with greater salt tolerance than barley, wheat, and corn [112–114], as well as vegetable crops, such as spinach, carrots, onion, and even asparagus [115]. A high variability in salinity tolerance among quinoa genotypes has been reported [20,116–119]. Traditionally, only genotypes from the Bolivian Salares were thought to have a high tolerance to salinity [12]. However, salinity tolerance in quinoa does not correlate with geographic distribution; varieties from coastal regions of Chile and highland areas outside the Salares ecoregion have similar or even higher salt tolerance levels [113,118,120–122]. Additionally, a wild relative of quinoa (*Chenopodium hircinum*) was found to have a much higher salinity tolerance level than quinoa cultivars [118].

In general, quinoa can tolerate moderate to high levels of salinity, ranging from a salt concentration of 150 mM NaCl (electrical conductivity ~15 dS m^{-1}) to as much as 750 mM NaCl (electrical conductivity ~75 dS m^{-1}) [120], which is greater than the salinity of seawater (>45 dS m^{-1}) [26]. In contrast, yields for glycophyte crops, such as wheat, rice, corn, and peas, start declining when the soil solution exceeds 40 mM NaCl (electrical conductivity ~4 dS m^{-1}) [112,114,123].

The optimal salinity conditions for quinoa growth are between 100 to 200 mM NaCl [123–125]. Plant seedling and germination stages are the most sensitive to salinity, even for halophytes [126,127]. Salt concentrations between 100 to 250 mM NaCl do not affect germination rates in most quinoa genotypes [95,117,125,128–132]. However, concentrations between 150 to 250 mM NaCl delay the onset of germination [117,120,130]. Changes in invertase activity and soluble sugar metabolism have also been detected during the quinoa germination process under saline stress [130,133]. At the seedling stage, the sugar concentration can increase or decrease, depending on the genotype, in both cotyledons and roots when grown under 200 to 400 mM NaCl.

The osmotic stress produced by a high salt concentration increases ABA production in roots and subsequent transport to the leaves as a signal to regulate stomatal conductance. The closure of stomata reduces water loss but also CO_2 uptake, thereby inhibiting photosynthesis [134]. Different experiments in semi-controlled conditions under salinity treatments have been conducted in quinoa to evaluate photosynthesis. In two quinoa varieties, salares variety 'Utusaya' and 'Titicaca', the CO_2 assimilation was reduced to 25% and 67%, respectively, when plants were grown at 400 mM NaCl [135]. In the altiplano variety 'Achachino', an increase of salinity from fresh water to 250 mM NaCl reduced the net assimilation rate of photosynthesis from 30 µmol CO_2 m^{-2} s^{-1} to 10 CO_2 µmol m^{-2} s^{-1} in a Photosynthetic Active Radiation (PAR) level of 1500 µmol m^{-2} s^{-1} [136]. Another experiment with the valley variety 'Hualhuas' showed a 70% reduction in the net photosynthetic rate when plants were grown under a salinity level of 500 mM NaCl [137]. Saline groundwater treatments with 100, 200, 300, and 400 mM NaCl were used in an experiment 'Titicaca'. Results showed that increasing the water salinity from 100 to 400 mM NaCl reduced the net assimilation rate of photosynthesis by 48% and seed yield by 72% [138]. On the other hand, the elevated atmospheric CO_2 (540 ppm) mitigated the effect of high salinity by tempering the stomatal limitation effect on photosynthesis, and consequently, reducing the hazard of oxidative stress [139].

The variety 'Titicaca', grown in field conditions under 22 dS m^{-1} and limited water in a Mediterranean environment, exhibited no yield reduction [44,51]. In another experiment with the same variety, the seed yield was reduced by 32% when plants were grown under 40 dS m^{-1} compared to the control (0 dS m^{-1}) [69]. However, under the same Mediterranean conditions, the sea-level variety 'Red head' grown under 30 dS m^{-1} exhibited a high susceptibility to salinity; various physiological parameters such as photosynthesis were affected [60].

Recently, new approaches such as halotolerant rhizobacteria and seed priming have been studied as alternatives to improve quinoa's physiological response to salinity stress [140,141]. For example, plant growth-promoting rhizobacteria have been used to alleviate the damage caused by salt stress because of their ability to fix nitrogen, produce siderophores, dissolve mineral insoluble phosphate, and produce phytohormones [142]. Seed priming partially hydrates seeds to the point of initiating the germination process. Treated seeds are then usually re-dried before planting. Different priming techniques are available, depending on which seed-embedding substance is used [143].

Yang et al. (2016) studied the relationship between plant growth-promoting halotolerant bacteria (*Enterobacter* sp. and *Bacillus* sp.) and quinoa under saline conditions [140]. Results showed that both strains mitigated the negative effects of salinity, reducing Na^+ uptake and improving water relations when the plants were grown at 300 mM NaCl. The same research group demonstrated that using saponin as a seed primer bio-stimulated quinoa germination under 400 mM NaCl [141]. In addition, seed priming 'Titicaca' with water (hydropriming) and with polyethylene glycol (osmopriming) showed that both hydropriming and osmopriming improved germination in salinity conditions [144].

Paclobutrazol, a gibberellic acid biosynthesis inhibitor, has been used to increase the yield and reduce plant height in quinoa [145]. More recently, Waqas et al. (2017) used this approach to mitigate salt stress in quinoa. They applied paclobutrazol (20 mg/L) on the leaves of sea-level variety 'Pichaman' under high salinity conditions (400 mM NaCl). Results showed improved chlorophyll and carotenoid content, enriched stomatal density on both leaf surfaces, and increased accumulation of osmoprotectants and antioxidants in leaf and root tissues [146]. All approaches described above could be excellent alternative tools to improve the quinoa yield in high salinity conditions.

3.2. Epidermal Bladder Cells and Stomatal Density

Morphological traits such as stomatal density and epidermal bladder cells (EBCs) have been studied in quinoa under salinity stress [119,120,136,147–151]. EBCs are modified epidermal hairs and classified as trichomes, along with glandular hairs, thorns, and surface glands. EBCs are shaped like gigantic balloons, with a diameter that is around 10 times bigger than epidermal cells, and can sequester 1000-fold more Na^+ compared with regular leaf cell vacuoles. EBCs accumulate water and different metabolites, such as betaine, malate, flavonoids, cysteine, inositol, pinitol, and calcium oxalate crystals. The main function of calcium oxalate is to regulate calcium levels and protect the plant against herbivores [152]. EBCs likely play a role in plant tolerance to ultraviolet (UV) light because they accumulate betacyanins and flavonoids, which are associated with UV protection and water homeostasis functions [153,154]. EBCs in *Chenopodium* species have a defensive function against herbivore insects and serve as structural and chemical components of defense [155]. EBCs play important roles in halophyte plants, such quinoa and Atriplex species. Their primary roles are the sequestration of sodium, improved K^+ retention, and the storage of metabolites, which help to modulate plant ionic relations, mainly gamma-aminobutyric acid [147].

In quinoa, EBCs are localized in leaves, stems, and inflorescences (Figure 2A,B). EBC density does not increase in response to high salinity [120,136]. However, the number of EBCs is greater in young leaves than in old leaves [135,156]. High sequestration of Na^+ in the EBCs from young leaves of quinoa under saline conditions (400 mM NaCl) has been reported [156]. Recently, with the assembly of a draft quinoa genome for the salares variety 'Real', transcriptome sequencing on bladder cells under both salt-treatment (100 mM NaCl) and non-treatment conditions has been conducted [148].

Stomatal area and density have been studied in quinoa under a variety of salinity conditions [120,136,151]. A salinity concentration of 400 mM NaCl was shown to reduce the number of stomata per leaf area in young, intermediate, and old leaves in 'Titicaca' [151]. Similar results were observed in the Chilean sea-level variety 'BO78'; the highest reduction in stomatal density (54%) was under 750 mM NaCl, compared with the untreated control [120]. Another study with 14 quinoa varieties by Shabala et al., (2013) demonstrated that stomatal densities in all varieties decreased when plants were grown in 400 mM NaCl [119]. Furthermore, the study found a strong positive correlation between stomatal density and plant salinity tolerance. Opposite results were observed in 'Achachino', where stomatal density increased ~18% when the plants were grown at 250 mM NaCl; nonetheless, stomatal size was reduced by the salinity effect [136]. Stomatal density and size could be a key mechanism for optimizing water-use efficiency under saline conditions.

Figure 2. Epidermal bladder cells (EBCs). (**A**) EBCs localized in the quinoa flower. Left, micrograph of quinoa flower. Right, scanning electron micrograph of quinoa flower. (**B**) Micrograph of quinoa leaves in two different varieties of EBCs.

3.3. Mechanisms of Salt Tolerance

Several studies to identify mechanisms of salt tolerance in quinoa have been conducted over the last 15 years. Quinoa has diverse mechanisms to withstand high levels of salt, including the efficient control of Na^+ sequestration in leaf vacuoles, xylem Na^+ loading, higher ROS tolerance, better K^+ retention, the upkeep of low cytosolic Na^+ levels, the reduction of slow and fast tonoplast channel activity, and a high rate of H^+ pumping in the mesophyll cell [26,33,156,157].

The accumulation of compatible solutes, such as proline and total phenolics, is associated with salt tolerance in quinoa [51,59,125,131,132,158,159]. However, Ismail et al., (2016) showed that proline may not play a major role in either osmotic adjustment or in the tissue tolerance mechanism [160]. But, the non-enzymatic antioxidant rutin improves quinoa salinity tolerance by scavenging hydroxyl radicals. Choline (Cho^+) is a metabolic precursor for glycine betaine and plays an important role in the osmotic adjustment to salinity stress in quinoa [161]. Polyamines were studied in four Chilean varieties under control (0 mM NaCl) and 300 mM NaCl conditions. The total amount of polyamines was reduced under the salinity conditions; however, the ratio of (sperdimidine+spermine)/putrescine increased up to 10-fold [132]. In addition, the activity of antioxidant enzymes changes in response to salinity in quinoa. For example, Panuccio et al. (2014) observed that, in quinoa seedlings of 'Titicaca', the regulation of antioxidant enzymes depended on the type and concentration of salt [129]. NaCl resulted in higher activity levels of superoxide dismutase (SOD), peroxidase (POX), ascorbate peroxidase (APX), and catalase (CAT), compared to the other salts evaluated (KCl, $CaCl_2$, $MgCl_2$). The exception was seawater from the Tyrrhenian Sea, which produced a major increase in POX, APX, and CAT activity [129].

High concentrations of NaCl can generate K^+ and H^+ fluxes in quinoa roots to the apoplast; thus, the activation of plasma membrane H^+-ATPase is needed to avoid further K^+ leakage from the cytosol [120,125]. H^+-ATPase is one of the active transporters, along with channels and co-transporters, that maintain intracellular K^+ and Na^+ homeostasis [162]. An analysis of cytosolic Na^+ showed that

Na$^+$ is removed quickly from the cytosol. The high K$^+$ concentration in both roots and shoots permitted higher pump activity when plants were grown under moderate salinity conditions [123]. Likewise, Cho$^+$ blocks the tonoplast slow vacuolar channels in quinoa leaf and root tissue, triggering efficient Na$^+$ sequestration [161].

Recently, the activity of mitogen-activated protein kinase (MAPK) under saline conditions (400 mM NaCl) was studied in quinoa seeds and seedlings. MAPK activities were highest in seeds and decreased during germination. Changes in MAPK activities occurred soon after imbibition, either with water or salt. Furthermore, under the salinity conditions, the decrease in MAPK activity occurred sooner than in non-stressed conditions [89].

3.4. Seed Composition Under Salinity Conditions

A few studies have reported changes in seed quality under saline conditions. The nutritional quality of 10 varieties of quinoa (nine from the Bolivian Altiplano and one from the Northwest Andean region of Argentina) was evaluated in two locations: Encalilla-Argentina (electric conductivity 2 dS m^{-1}) and Patacamaya-Bolivia (electric conductivity 7 dS m^{-1}). Results demonstrated that essential amino acids were affected more by salinity than yield and protein level. Seven of the 10 varieties studied exhibited increased essential amino acids when grown in the higher salinity location [91].

Quinoa varieties 'Titicaca' and 'Q52' were evaluated under field conditions in Italy using different irrigation regimes under saline conditions (22 dS m^{-1}). Results showed that seed fiber and saponin content decrease in response to the highest level of saline water; however, protein content was unaltered [44,163]. Similar results in seed fiber were found in the variety 'Hualhuas' under field conditions of 17.9 dS m^{-1} in the northwestern part of Sinai-Egypt [124]. On the other hand, seed protein content in eight different varieties increased under a saline-sodic soil (6.5 dS m^{-1}) in Larissa-Greece [164]. Similarly, four sea-level quinoa varieties, consisting of 'CO407D (PI 596293)', 'UDEC-1 (PI 634923)', 'Baer (PI 634918)', and 'QQ065 (PI 614880)', exhibited increased seed protein content when grown under 32 dS m^{-1} Na$_2$SO$_4$ conditions. In contrast, the study found no change under the same concentration of NaCl [165].

Prado et al. (2014) found variation in the concentrations and tissue distributions of 18 mineral elements in the seeds of seven different quinoa cultivars from Patacamaya, Bolivia (3960 m above sea level) and from Encalilla, Argentina (1980 m above sea level) [92]. The data clearly showed inter- and intra-varietal differences in seed mineral concentrations between the two sites, strongly suggesting that G × E interactions were responsible for mineral variation among the quinoa cultivars. In another study, the mineral content of calcium (Ca), magnesium (Mg), zinc (Zn), and manganese (Mn) in quinoa seeds decreased in response to saline-sodic soil in Larissa-Greece [164]. Similar results were found for the valley variety 'Hualhuas' under field conditions, with 17 dS m^{-1} in the northwestern part of Sinai-Egypt [124]. Using X-ray microanalysis, the study found high Na$^+$ accumulation in the pericarp and embryo tissues of quinoa seeds, but low amounts in the perisperm. Furthermore, concentrations of essential minerals such as Fe increased due to the high salinity conditions [124].

Proteomic and amino acid profiles, phenolic content, and antioxidant activity of protein extracts of seeds from three quinoa varieties, consisting of salares variety 'R49' and sea-level varieties 'VI-1' and 'Villarrica', were analyzed under two salinity levels (100 and 300 mM NaCl) (Aloisi et al., 2016). Results showed a reduction in all amino acids derived from protein hydrolysis in 'VI-1' and 'Villarrica'. However, several amino acids remained unchanged or increased with increasing salinity in 'R49'. Total polyphenol content increased in the three genotypes with increasing salinity, with the largest increase in the ecoytpe 'R49'. Similarly, the increases in total flavonoids and total antioxidant activity were more evident in 'R49' [116]. In contrast, no change was observed in total polyphenol content in the sprouts of 'B080' under 150 mM NaCl conditions; nevertheless, sprout growth was reduced [95].

3.5. Gene Expression Under Saline Conditions

Ruiz et al. (2016) described important transcription genes related to the salinity response in quinoa [33]. However, with the recent publication of one robust and two draft quinoa genomes [34,148,166], several new potential genes have been identified that may also play a role in quinoa's response to salt stress. Table S1 summarizes the genes and candidate genes that have been studied in quinoa under saline conditions.

Exclusion of Na^+ from the cytoplasm is encoded primarily by two genes. One gene is *Salt Overly Sensitive 1* (SOS1), which encodes a Na^+/H^+ antiport located at the plasma membrane of epidermal root cells and functions to extrude Na^+ out of the cell [167–169]. The other gene is tonoplast-localized Na^+/H^+ exchanger 1 (NHX1), which sequesters Na^+ inside the vacuole [170,171]. Maughan et al. (2009) cloned, sequenced, and characterized two homoeologous *SOS1* loci (*cqSOS1A* and *cqSOS1B*) in saline conditions (300 mM NaCl) using the quinoa salares variety 'Ollague' [172]. They observed that both genes were up-regulated in the leaves, but not in the roots. Similar results were reported for other quinoa varieties when plants were grown in 300 mM NaCl and 450 mM NaCl [132,173]. However, up-regulation of the gene CqNHX1in shoots and roots was observed in a salt-tolerant variety from Chile when plants were grown in 300 mM NaCl [33,122,132]. In addition, transcription levels of tonoplast intrinsic protein 2 (TIP2) and betaine aldehyde dehydrogenase (BADH) were reported when the two salares varieties 'Ollague' and 'Chipaya', and one valley variety 'CICA', were grown in 450 mM NaCl [173]. Up-regulation of BADH in roots was found in the salares-type genotypes, indicating that betaine plays an important role in decreasing salt stress in roots. Moreover, results revealed that other genes are involved in the mechanisms of the salt stress response [173].

Abscisic acid (ABA), polyamine (PA), and proline biosynthesis genes were studied in two quinoa varieties, 'R49' from salares and 'Villarica' from sea-level under saline conditions (300 mM NaCl). The expression of 22 genes was common to both varieties. The salt adaptation mechanism was based primarily on ABA-related responses. For example, the gene encoding for the key enzyme in ABA biosynthesis 9-*cis*-epoxycarotenoid dioxygenase (NCED) was the most strongly induced [122]. Likewise, a phylogenetic analysis showed that gene families in ABA signaling were distributed more often in the quinoa genome compared to other *Amaranthaceae* species [166]. Identification of ortholog genes involved in ABA biosynthesis, transport, and perception in quinoa under saline conditions was reported. Hence, quinoa contains neoxanthin synthase (NSY), ABA4, short-chain dehydrogenases/reductases (SDRs) genes, and 11 NCEDs, which are involved in the ABA biosynthetic pathway. The number of these genes in quinoa is nearly two-fold compared to other diploid plants. For example, a higher number of ABA receptor and transportation genes was observed in quinoa, with 22 ABA receptor pyrabactin resistant (PYL) family genes and 81 genes from the ABC transports group (ABCGs) compared to 10 PYL and 34 ABCGs in *Amaranthus hypochondriacus* [148].

A transcriptome analysis of bladder cells in quinoa compared a salinity treatment (100 mM NaCl) to non-treated conditions [148]. Results showed a higher expression of genes relative to energy import and ABA biosynthesis in bladder cells compared with the leaf lamina. For instance, anion transporter genes, such as cell anion channels (SLAH), nitrate transporter (NRT), and chloride channel protein (C1C), and cation transporter genes, including NHX1 and K^+ transporter (HKT1), exhibited a higher expression in bladder cells. After the salt treatment, 180 and 525 differentially expressed genes were identified in leaf lamina and bladder cells, respectively. However, the two tissues shared only 25 genes, indicating that leaf and bladder cells respond differently to salinity [148]. Additionally, genes involved in suberin and cutin biosynthesis are significantly enriched in bladder cells under salinity. On the other hand, photosynthesis and chloroplast protein-encoding genes were down-regulated. The transcript levels of two NCED genes and some of the short-chain SDR genes in bladder cells were six-fold and 1000-fold higher, respectively, than in leaf cells. Furthermore, an elevated expression of ABA transporter and ABA receptor genes was found in bladder cells. Together, the above results suggest that bladder cells might maintain a high level of ABA homeostasis [148]. The ABA biosynthesis pathway shares responsive neoxanthin with an upregulation of NCED genes from both drought and salt stress [148].

RNA-seq analyses with a comparative genomics and topology prediction approach were conducted to identify new transmembrane domain genes in quinoa under saline conditions (300 mM NaCl); 1413 genes were differentially expressed in response to salt [118]. However, 219 genes were chosen after selecting for only genes that encoded proteins with more than one predicted transmembrane domain. These 219 candidate genes were further studied using the sequence information of 14 quinoa varieties (six sea-level, four altiplano, two valley, and two salares), five *C. berlandieri* accessions, and two *C. hircinum* accessions [34], and physiological data under salinity conditions. Using copy number variation (CNV) and the presence of SNPs between the five most salt-tolerant and the five most salt-sensitive accessions, 14 candidate genes were identified, and six SNPs were located in the first exon of the *AUR62043583* gene (Table S1). Thus, the study found 15 new candidate genes that could contribute to the differences in salinity tolerance among quinoa varieties [118].

Betalains are tyrosine-derived, red-violet and yellow pigments found exclusively in *Caryopyllales* plants, including quinoa. Betalains are involved in salt stress tolerance due to their antioxidant activity [174]. Mutagenesis using ethyl melthanesulfonate on the quinoa variety 'CQ127' revealed that the gene *CqCYP76AD1-1* is involved in the green hypocotyl mutant [175]. This gene was then isolated and shown to be light-dependent in quinoa hypocotyl. These findings suggest that *CqCYP76AD1-1* is involved in betalain biosynthesis during the hypocotyl pigmentation process in quinoa [175]. This gene should be interesting to study under salinity stress because the betalain accumulation could play an important role in protecting quinoa hypocotyl.

In conclusion, due to its recently sequenced genome and high tolerance to salt stress, quinoa has become an important model crop to further our understanding of how plants respond to salinity. In recent years, using the new molecular tools, several novel genes have been reported. However, validation of these genes is necessary; thus, efforts to transform quinoa have begun to understand the functions of these genes. Furthermore, quinoa's strong genotype-dependent response to salinity offers breeders the opportunity to work with diverse quinoa genotypes to develop new salt-tolerant varieties with a high grain quality and other valuable traits.

4. High Temperature

Excessively high temperature during plant growth is considered one of the most important abiotic stresses, and is being reported with increasing frequency due to the consequences of present-day climate change [176]. Worldwide, extensive agricultural losses have been attributed to heat, often in combination with drought [1,8,177]. Heat stress in plants is defined as an increase in air temperature above the optimum growth temperature for a length of time sufficient to cause damage and, hence, limit growth and development [178]. Heat stress produces different responses across plant species, depending on the temperature duration and the plant developmental stage [179,180].

Effects of heat stress include: (1) morphological changes, such as the inhibition of shoot and root growth and increased stem branching; (2) anatomical changes, such as reduced cell size and increased stomatal and trichome densities; and (3) phenological changes [5,178,181]. In addition to morpho-anatomical changes, physiological effects of heat stress include protein denaturation; increased membrane fluidity; cytoskeleton instability; changes in the respiration, photosynthesis, and activity of carbon metabolism enzymes; osmolyte accumulation; chloroplast and mitochondrial enzyme inactivation; changes in phytohormones, including ABA, salicylic acid, and ethylene; and the induction of secondary metabolites [178,182].

Heat stress induces oxidative-stress-generating ROS, in the same way than in drought or salinity stress, high accumulations of ROS cause serious plant toxicity; however, low concentrations of ROS act as a signaling molecule that activates other plant processes, such as programmed cell death [5,176,178]. Finally, heat-shock proteins (HSPs) play a central role in the heat stress response (HSR) when plants suffer from either an abrupt or gradual increase in temperature [178,183]. Several studies have indicated the importance of HSPs in thermotolerance in many plant species; hence, HSP70 and HSP90 are indispensable to induce thermotolerance [183,184]. Heat stress factors (HSFs) serve as the terminal component of signal transduction of HSP expression. [183,184]. In Figure 3, we show the primary physiological responses to drought, salinity, and heat in quinoa.

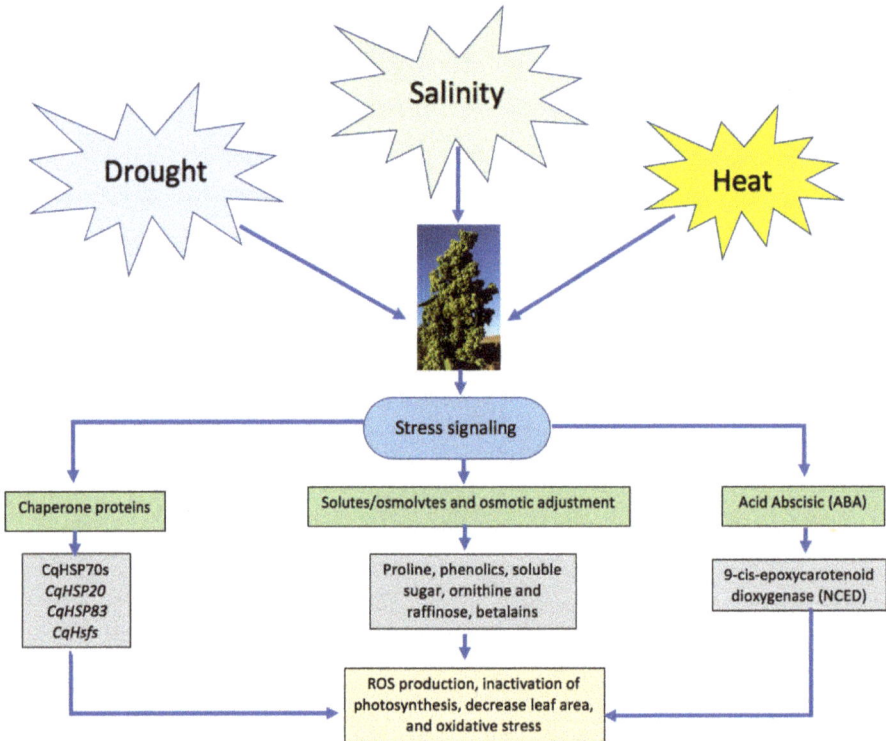

Figure 3. Primary physiological responses to drought, heat, and salinity on quinoa. The expression of NCED genes is upregulated to drought and salinity and chaperone proteins are upregulated to drought and heat stress.

Quinoa can tolerate a wide range of temperatures (from −8 °C to 35 °C) and relative humidity conditions (from 40% to 88%), depending on genotype characteristics and phenological stage [185]. Despite the adaptation of quinoa outside the Andes [24,42,43,124,186–192], a high temperature during flowering and seed set can significantly reduce the yield and is one of the major barriers to the global expansion of quinoa. For example, studies in Italy [45], Morocco [193], Germany [194], Portugal [195], India [27], Egypt [124], Mauritania [42], and the United States [71,196] have reported that high temperatures reduce the quinoa seed yield. Most research has focused on understanding the effect of temperature on quinoa seed germination. However, few studies have focused on understanding the physiological changes in quinoa under high temperature during other phenological stages.

4.1. High Temperature Effects on Quinoa Germination

Numerous studies have been carried out in different temperature regimes to describe the effect of temperature on quinoa germination [117,197–205]. For instance, studies have found a positive linear relationship between the germination rate and temperature in quinoa [199,202,205]. Findings suggested that the optimal germination temperature is 30 to 35 °C, maximal germination temperature is 50 °C, and base germination temperature is 3 °C [199,202]. In contrast, Bois et al. (2006) reported that the base germination temperatures for 10 different quinoa varieties varied between −1.9 and 0.2 °C [205]. Another study with the variety 'Titicaca', and salares varieties 'Santa Maria', and 'Sajama', used three different models to show that the optimum germination temperature range was 18−36 °C for Sajama and 22−35 °C for the other two varieties [203]. The study also reported a base germination temperature of 1.0 °C and a maximal germination temperature of 54.0 °C for the three varieties evaluated. Quinoa seeds can be stored up to 430 days under controlled environmental conditions, at a constant temperature of 25 °C, before germination completely declines [204]. According to these results, quinoa is highly heat tolerant during its germination stage, and is capable of germinating under a wide range of temperatures, from very low (−1.9 °C) to very hot (>48.0 °C).

4.2. High Temperature Effects on Quinoa Growth and Physiological Parameters

The base temperature (T_b) threshold for quinoa development is variable; for example, in flowering time and leaf appearance, T_b is 1 °C, whereas in leaf width, the T_b increased to 6 °C [205]. T_b can change due to different development rates and the latitude origin of the genotypes [206]; for instance, T_b for sea-level variety 'Baer I' is 6.4 °C, whereas T_b for valley variety 'Amarilla de Marangani' is 3.7 °C [207].

Temperatures above 35 °C in the flowering and seed fill stages have been associated with significant reductions in yield. At these temperatures in quinoa fields near Pullman, WA, Peterson and Murphy (2015b) and Walters et al. (2016) observed that inflorescences either lacked seeds or contained empty seeds when the temperature increased above 35 °C [71,196]. Similarly, Bonifacio (1995) observed both the reabsorption of seed endosperm and inhibition of anther dehiscence in quinoa flowers due to high temperature (35 °C) at the flowering stage [208]. However, varietal differences in heat tolerance have been detected in quinoa. For example, sea-level varieties 'Colorado 407D', 'QQ74', and 'Kaslaea' showed greater heat tolerance under field conditions in Pullman compared to other sea-level varieties grown in the same conditions [196].

High night temperatures were evaluated in one commercial sea-level variety, 'Regalona', and one quinoa landrace, 'BO5', under field conditions in Chile. Results showed that night temperatures between 20–22 °C (~4 °C above the night ambient air temperature) during the flowering stage reduced the seed yield by between 23–31% and negatively affected the biomass and number of seeds. On the other hand, seed protein and harvest index were unaffected [209].

High temperature in quinoa has also been studied in combination with other stresses, such as drought, high salinity, and elevated CO_2 [54,136,210]. A controlled experiment with 'Titicaca' was conducted under cool temperatures of 18/8 °C (day/night temperature) and high temperatures of 25/20 °C, with three different irrigation regimens, consisting of full irrigation, deficit irrigation, and partial root-zone drying [54]. Results showed that drought has a major negative effect on physiological parameters compared to high temperature. In contrast, high temperature increased stomatal conductance, leaf photosynthesis, leaf chlorophyll fluorescence, and chlorophyll content index. On the other hand, anions and cations from the xylem sap increased in response to high temperature, showing that quinoa can adjust osmotically to overcome increased transpirational water. Similar results were observed in stomatal conductance and leaf photosynthesis in altiplano variety 'Achachino', grown at 28/20 °C, and sea-level varieties 'QQ74 (PI 614886)' and '17GR (AMES 13735)', grown at 40/24 °C [136,211].

The altiplano variety 'Achachino' was evaluated under high temperature 28/20 °C. Results showed that plant dry mass and yield were unaffected by high temperature; however, more and longer branches were observed in plants due to high temperature [136]. Similar responses were described with quinoa sea-level varieties 'QQ74' and '17GR' under 40/24 °C (Figure 4) [211].

Figure 4. Impact of high temperature on the length of the secondary axis from the quinoa inflorescence. Plant grown under control conditions 22/16 °C (**left**) and high temperature 40/24 °C (**right**).

Bunce (2017) studied the effect of high temperature and high concentration of CO_2 in two quinoa sea-level varieties, 'Red Head' and 'Cherry Vanilla', and one altiplano variety, 'Salcedo', during the anthesis stage [210]. Results showed that the harvest index in all varieties either increased or remained the same in response to high temperature (35/29 °C). Seed dry mass decreased in 'Cherry Vanilla' when grown at the high temperature and under the ambient CO_2 concentration (400 µmol·mol^{-1}). However, for the other two varieties, when grown under high temperature and either ambient or high CO_2 concentrations (600 µmol·mol^{-1}), seed dry mass was higher than or the same as the control conditions (20/14 °C). Recently, another study grew 'Cherry Vanilla' under cool (12/6 °C, 20/14 °C) and moderate (28/22 °C) temperatures. Results showed that this quinoa variety has a large capacity for thermal acclimation to temperature, depending on the maximum carboxylation capacity of Rubisco [212]. However, the effect of high temperature also depends on the variety origin. For example, valley variety 'Amarilla de Marangani' produced its heaviest seeds at a day temperature of 20 °C; whereas, quinoa from the sea-level variety, 'NL-6', produced its heaviest seeds at a day temperature of 30 °C [186]. Membrane stability of quinoa leaves was measured in six quinoa varieties when grown at 34/32 °C (acclimated) and 22/20 °C (non-acclimated). Results showed that quinoa altiplano variety 'Illpa' exhibited less cellular damage after its leaves were exposed to 50 °C for 64 min under acclimated conditions compared to the same exposure under non-acclimated conditions [213].

Quinoa pollen was evaluated for two sea-level varieties, 'QQ74' and '17GR', under high temperature (40/24 °C) in growth chamber experiments. Results demonstrated that quinoa pollen viability decreased, but without a concomitant effect on seed set and no morphological changes in the pollen surface (Figure 5). The latter finding was probably due to the high amount of pollen produced by the plants and the high relative humidity (40–65%) recorded in the in the growth chambers [211]. However, field experiments in Pullman, WA-USA, using eight quinoa varieties recorded reduced or total losses in the seed yield, most likely due to the low relative humidity (less than 30%) combined with high temperature (35 °C) and pest pressure.

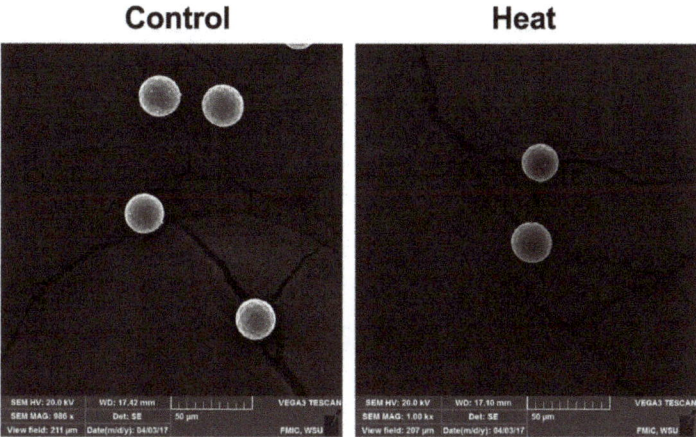

Figure 5. Scanning electron micrographs of quinoa pollen under control conditions 25/16 °C and heat conditions 40/24 °C, magnification 1000×

The heat shock transcription factor (*Hsf*) gene family was studied in quinoa; 23 (*CqHsfs*) genes were identified. The expression profiles from *CqHSFs* genes were explored using RNA-seq data. Four CqHsfs upregulated in the expression profile were then validated in the salar ecotype 'Real-Blanca' under high temperature (37 °C), and the results showed that *CqHsfs3* and *CqHsfs9* had a higher expression level after 6 h of heat treatment, while *CqHsfs4* and *CqHsfs10* showed a higher expression level at 12 h [214].

4.3. Photoperiod and Temperature Effects on Quinoa

In latitudes greater than 30°, where temperatures exceed 30 °C during the growing season and photoperiods exceed 14 h, quinoa varieties from Andean valleys are low yielding [71,196,215]. In a study with altiplano variety 'Kanckolla', the seed diameter decreased as much as 73% when the air temperature rose to 28 °C on long days (16 h) compared to the seeds of quinoa plants grown at 21 °C on short days (10.25 h) [215]. Two models were used to quantify photoperiod and temperature responses in nine short-day varieties from emergence to visible flower buds. Results showed that both models were similar in their goodness of fit. Photoperiod and temperature parameters were not significantly related to latitude of origin; however, a negative association was observed when the attributes evaluated were considered as constants [216]. For the nine varieties evaluated in Bertero et al. (1999b), both temperature and photoperiod controlled the rate of leaf appearance [215]. Temperature sensitivity was the highest for quinoa varieties originating in cold or dry climates, whereas temperature sensitivity was the lowest for varieties from humid and warmer climates [217]. Solar radiation affects phyllochron in quinoa; thus, varieties from Peru, Bolivia, and Southern Chile are more sensitive to the radiation than Ecuadorian varieties. However, Ecuadorian quinoas are highly sensitive to photoperiod and exhibit the longest phyllochron [218]. Saponin content in sea-level varieties 'Regalona' and 'Roja', and valley variety 'Tunkahuan', was evaluated under short days and long days (8 h and 16 h, respectively) and under two temperatures (20 °C and 30 °C) [219]. Results showed that the highest saponin content occurred in plants grown under short days and 30 °C.

In conclusion, quinoa cultivation has expanded worldwide as a crop because of its capacity to thrive in high temperature environments. Although many studies report that plant growth and development are not limited by high air temperatures (~40 °C), reproductive stages can be affected; for example, quinoa pollen viability is reduced at 40 °C. Nevertheless, more studies under both field and semi-control conditions are necessary to evaluate the interactions between high temperature and other abiotic and biotic stressors. Additional studies are also needed that evaluate more genotypes with different planting dates, under tropical and sub-tropical environments, and at different altitudes. Furthermore, wild *Chenopodium* species, such as *C. berlandieri* and *C. hircinum*, are adapted to high temperatures in their native habitats. Thus, another option to develop enhanced heat tolerance in quinoa is interspecific and intergeneric hybridization with these species [220–223].

5. Ultraviolet B (UV-B) radiation

Ultraviolet B (UV-B) radiation represents a small fraction of the solar spectrum (280–315 nm); however, its high energy can be harmful to living organisms [224]. Plants respond differently to UV-B based on their age [225], species origin [226], or circadian rhythms [227]. Studies about UV RESISTANCE LOCUS8 (UVR8) photoreceptor opened the query of whether UV-B should be considered as an abiotic stress or a morphogenetic factor in crop production [228–230].

The UV-B effect on quinoa has been mainly studied in South American countries at high altitudes, where UV-B exposure is higher than in other parts of the world [231]. Palenque et al. (1997) reported different responses in morphological and pigment synthesis across quinoa altiplano varieties 'Chucapaca', 'Robura', and 'Sayaña' [232]. They found an increase in leaf flavonoid content and a reduction in plant height and quinoa leaf size in the treatment directly exposed to UV-B; however, 'Chucapaca' exhibited the best adaptation to UV-B. Sircelj et al. (2002) revealed the effects of UV-B at metabolic and ultrastructural levels in quinoa seedlings. For instance, the thylakoid organization changed in response to exposure to UV-B [233]. Hilal et al. (2004) found that epidermal lignin deposition in quinoa's cotyledons was induced by UV-B radiation [234]. Additionally, González et al. (2009) used a semi-control experiment with altiplano varieties 'Chucapaca' and 'Cristalina' to study the quinoa response to different levels of UV-B [235]. Results demonstrated that sucrose, glucose, and fructose exhibited different distribution patterns in cotyledons and leaves of both varieties, depending on whether exposure was to near-ambient or strongly reduced UV-B. These studies are useful to understand the plasticity of metabolic pathways involved in a plant's tolerance to solar UV-B radiation.

Another study, under controlled conditions, showed changes in morphological responses, such as plant height, stem diameter, leaf number, and specific leaf area, in different quinoa varieties due to UV-B radiation [236]. The effects of UV-B radiation on photosynthetic (total chlorophyll, chlorophyll *a*, chlorophyll *b*, and carotenoids) and protective (UV-B absorbing compounds) pigments and soluble sugars (glucose, fructose and sucrose) in five quinoa varieties from different geographic origins were studied by Prado et al., (2016) [237]. A common response observed across the five varieties was an increase in the content of UV-B absorbing compounds, which showed a high peak in the absorbance region of 305 nm. The researchers proposed that these compounds act as a "chemical shield" that protect a plant's photosynthetic apparatus against the excess energy from radiation exposure.

Recently, Reyes et al. (2018) reported the first study on the effect of UV-B on quinoa photosynthesis [238]. In this study, chlorophyll fluorescence, pigment synthesis, photosynthesis, and ROS accumulation were affected by different levels and duration of UV-B. In summary, quinoa can regulate different mechanisms of response, depending on the UV-B irradiation dosage. Despite the progress in our understanding of the effects of UV-B on plants, additional studies are necessary to: (1) determine the UV-B threshold, where exposure ceases to be a natural morphogenetic factor and becomes a stress factor; and (2) clarify the relationship of UV-B levels with other natural environmental factors.

6. Frost, Waterlogging, and Heavy Metals

Other abiotic stressors, such as frost, waterlogging, and heavy metals, have been studied in quinoa [58,185,239–241]. Jacobsen et al. (2005) studied different quinoa genotypes under frost temperatures [185]. Results showed that varieties from the Altiplano of Peru tolerated −8 °C for 4 h during the two-leaf stage much better than varieties from Andean valleys, which are more sensitive to frost. For instance, altiplano varieties 'Witulla' and 'Ayara' experienced a 4.17% plant death rate under −8 °C for 4 h; whereas, valley varieties 'Quillahuaman' experienced a 25% death rate under

−8 °C for 4 h and 50% under −8 °C for 6 h. Furthermore, the flowering stage is even more sensitive to frost; the study recorded yield reductions of 56% in the variety 'Quillahuaman' and 26% in the variety 'Witulla' when plants were exposed to −4 °C for 4 h. Another study with 'Witulla' and 'Quillahuaman' demonstrated that the main quinoa response mechanism to frost is avoidance of ice formation, which is facilitated by the plant's high soluble sugar content in 'Witulla'. Thus, proline and soluble sugar contents, such sucrose, could be used as an indicator of frost resistance [240]. Similar results were observed in the altiplano variety Sajama grown under 5/5 °C, where the low temperature induced sucrose-starch partitioning in quinoa cotyledons [242]. Moreover, a low temperature might induce different regulatory mechanisms linked to changes in invertase, sucrose synthase, and sucrose-6-phosphate synthase activity in cotyledons and embryonic axes during Sajama seed development [133].

An experiment in controlled growth chambers with the altiplano variety 'Sajama' showed that waterlogging produced several negative effects, including: (1) decreased plant and root dry weights; (2) low total chlorophyll, chlorophyll *a*, and chlorophyll *b* contents; and (3) high amounts of soluble sugars and starch [58]. Under field conditions in Brazil, the variety 'BRS Piabiru' exhibited maximum leaf measurement values when plants were grown in a water regimen of 563 mm. However, under 647 mm, reductions in leaf measurements were observed, indicating the sensitivity of quinoa to excess water [241]. Pre-harvest sprouting in quinoa could be a serious problem in places with high precipitation, especially when the rain coincides with the seed-set stage. The quinoa sea-level variety 'Chadmo, QQ065-PI 614880' originating from the humid area of Chiloe island in Chile, may be a good choice for these conditions because of its demonstrated higher seed dormancy and greater pre-harvest sprouting tolerance [243,244]. In a comparative study in the high precipitation area of the Olympia Peninsula in Washington State, USA, 'Chadmo' showed a higher level of pre-harvest sprouting tolerance [196].

Bhargava et al. (2008b, 2008a) studied different *Chenopodium* spp. under heavy metal soil conditions [3,239]. Results showed that 17 quinoa accessions accumulated high amounts of most heavy metals, such as zinc (Zn), chromium (Cr), nickel (Ni), and cadmium (Cd), in their leaves compared to the other species. For instance, the accessions *C. quinoa* PI 587173, *C. quinoa* PI 478410, *C. quinoa* Ames 22158, and *C. giganteum* CHEN 86/85 accumulated the highest contents of Cd; *C. quinoa* PI 510,536 and Ames 22,156 accumulated the highest contents of Ni, Cr, and Zn. Furthermore, another study in a contaminated urban area 'brownfield' in Vancouver, Canada, showed that quinoa from the altiplano variety 'Quinoa de Quiaca—PI 510532' is a hyperaccumulator of heavy metals such as Cd, copper (Cu), and lead (Pb). Consequently, quinoa seeds would be inappropriate for human consumption due to high concentrations of trace metals if grown in brownfield areas [245]. A study of quinoa's physiological response to various concentrations of Cr showed that leaves from sea-level variety 'Regalona' tolerated up to 1 mM external chromium(III) chloride ($CrCl_3$), activating tocopherol accumulation and enhanced tyrosine aminotransferase content. However, the highest doses of 5 mM Cr(III) produced oxidative stress, generating high hydrogen peroxide and proline contents [246]. Newly discovered heavy metal-related genes involved in the different mechanisms of accumulator plant species and recently published genetic resources could greatly advance novel non-consumptive functions for quinoa. For example, identifying promising functional molecular tools in chenopod species may lead to the effective exploitation of quinoa cultivation as a phytoremediation strategy for environmental contamination cleanup.

7. Conclusions

The physiological, biochemical, and morphological responses of different quinoa varieties to various abiotic stressors, under both field and lab conditions, show that quinoa has a wide plasticity and tolerance to those stressors. This tolerance and plasticity seem to be controlled genetically, and significant advances in breeding have been initiated with the whole genome sequencing of quinoa and the use of new molecular tools. One of the most relevant aspects of quinoa is its high salinity tolerance, unlike other crop species, such as wheat, rice, barley, and maize. Every year, arable land is lost due to salinization, extreme temperatures, and severe drought, which are being reported more often. Hence, farmers have begun to look for halophytic- and abiotic-tolerant species, such as quinoa that can perform under these conditions. Another main feature of quinoa is its high nutritional value, such as essential amino acids and mineral concentrations, which are maintained in spite of abiotic stress conditions. Additionally, quinoa could be considered a multipurpose plant, considering that seeds and leaves can be use as food, biomass can be used as animal feed or a cover crop, and plantings can serve as a phytoremediation tool for environmental cleanup.

Despite the numerous recent studies about abiotic stress on quinoa, much information remains unknown. Future studies should focus on the genetic underpinnings and mechanisms involved in how quinoa's abiotic stress tolerance influences its chemical composition. This additional information will allow quinoa breeders to generate new varieties that are widely adapted to a variety of environmental conditions, and in turn, facilitate quinoa's worldwide expansion. Likewise, the recent exploration of intercrosses between quinoa and its wild relatives should provide new genetic combinations with promising opportunities to breed for production in extreme conditions. Taken together, quinoa represents an excellent model to fully explore abiotic stress tolerance mechanisms and new genes to improve plants.

Supplementary Materials: The following is available online at http://www.mdpi.com/2223-7747/7/4/106/s1, Table S1: Candidate genes involved in the salinity tolerance in quinoa.

Author Contributions: L.H. and K.M. conceived and wrote the manuscript. F.B-M., J.G., and F.F. edited the manuscript and provided significant additional information. All authors read and approved the final manuscript.

Funding: This research was funded by The National Secretary of Science, Higher Education, Innovation and Technology of Ecuador (SENESCYT-Ecuador) and USDA NIFA Awards: 2016-51300-25808 and 2016-68004-24770.

Acknowledgments: We thank Washington State University Sustainable Seed Systems Lab (http://www.sustainableseedsystems.org/) for providing tools, guidance, and supervision for this work. Additionally, we thank Karen A. Sanguinet, Kimberly Garland Campbell, and Kulvinder S. Gill for their comments and Linda R. Klein for editing the final manuscript.

Conflicts of Interest: The authors declare no conflict of interest. The funders had no role in the design of the study; in the collection, analyses, or interpretation of data; in the writing of the manuscript, or in the decision to publish the results.

References

1. Mittler, R. Abiotic stress, the field environment and stress combination. *Trends Plant Sci.* **2006**, *11*, 15–19. [CrossRef] [PubMed]
2. Barnabás, B.; Jäger, K.; Fehér, A. The effect of drought and heat stress on reproductive processes in cereals. *Plant Cell Environ.* **2008**, *31*, 11–38. [CrossRef] [PubMed]
3. Bhargava, A.; Shukla, S.; Srivastava, J.; Singh, N.; Ohri, D. *Chenopodium*: A prospective plant for phytoextraction. *Acta Physiol. Plant.* **2008**, *1*, 111–120. [CrossRef]
4. Bhargava, A.; Carmona, F.F.; Bhargava, M.; Srivastava, S. Approaches for enhanced phytoextraction of heavy metals. *J. Environ. Manag.* **2012**, *105*, 103–120. [CrossRef] [PubMed]
5. Hasanuzzaman, M.; Nahar, K.; Alam, M.M.; Roychowdhury, R.; Fujita, M. Physiological, biochemical, and molecular mechanisms of heat stress tolerance in plants. *Int. J. Mol. Sci.* **2013**, *14*, 9643–9684. [CrossRef] [PubMed]
6. Munns, R.; Tester, M. Mechanisms of salinity tolerance. *Annu. Rev. Plant Biol.* **2008**, *59*, 651–681. [CrossRef] [PubMed]
7. Pereira, A. Plant abiotic stress challenges from the changing environment. *Front. Plant Sci.* **2016**, *7*. [CrossRef] [PubMed]
8. Suzuki, N.; Rivero, R.M.; Shulaev, V.; Blumwald, E.; Mittler, R. Abiotic and biotic stress combinations. *New Phytol.* **2014**, *203*, 32–43. [CrossRef] [PubMed]
9. IPCC. *Climate Change 2014—Impacts, Adaptation, and Vulnerability: Regional Aspects*; Cambridge University Press: Cambridge, UK; New York, NY, USA, 2014; ISBN 978-1-107-05816-3.

10. Meehl, G.A.; Tebaldi, C. More intense, more frequent, and longer lasting heat waves in the 21st Century. *Science* **2004**, *305*, 994–997. [CrossRef] [PubMed]
11. Xia, J.; Tu, K.; Yan, Z.; Qi, Y. The super-heat wave in eastern China during July–August 2013: A perspective of climate change. *Int. J. Climatol.* **2016**, *36*, 1291–1298. [CrossRef]
12. Tapia, M. The Long Journey of Quinoa: Who wrote its history. In *State of the Art Report on Quinoa around the World 2013*; Bazile, D., Bertero, H.D., Nieto, C., Eds.; FAO: Santiago, Chile; CIRAD: Montpellier, France, 2015; Volume 1, pp. 1–7, ISBN 978-92-5-108558-5.
13. Costa Tártara, S.M.; Manifesto, M.M.; Bramardi, S.J.; Bertero, H.D. Genetic structure in cultivated quinoa (*Chenopodium quinoa* Willd.), a reflection of landscape structure in Northwest Argentina. *Conserv. Genet.* **2012**, *13*, 1027–1038. [CrossRef]
14. Zurita-Silva, A.; Fuentes, F.; Zamora, P.; Jacobsen, S.-E.; Schwember, A.R. Breeding quinoa (*Chenopodium quinoa* Willd.): Potential and perspectives. *Mol. Breed.* **2014**, *34*, 13–30. [CrossRef]
15. Dillehay, T.D.; Rossen, J.; Andres, T.C.; Williams, D.E. Preceramic adoption of peanut, squash, and cotton in Northern Peru. *Science* **2007**, *316*, 1890–1893. [CrossRef] [PubMed]
16. Bertero, H.D.; De la Vega, A.J.; Correa, G.; Jacobsen, S.E.; Mujica, A. Genotype and genotype-by-environment interaction effects for grain yield and grain size of quinoa (*Chenopodium quinoa* Willd.) as revealed by pattern analysis of international multi-environment trials. *Field Crop. Res.* **2004**, *89*, 299–318. [CrossRef]
17. Bhargava, A.; Shukla, S.; Rajan, S.; Ohri, D. Genetic diversity for morphological and quality traits in quinoa (*Chenopodium quinoa* Willd.) germplasm. *Genet. Resour. Crop Evol.* **2007**, *54*, 167–173. [CrossRef]
18. Bhargava, A.; Ohri, D. Origin of genetic variability and improvement of quinoa (*Chenopodium quinoa* Willd.). In *Gene Pool Diversity and Crop Improvement*; Sustainable Development and Biodiversity; Rajpal, V., Rao, S., Raina, S., Eds.; Springer: Cham, Switzerland, 2016; pp. 241–270, ISBN 978-3-319-27094-4.
19. Fuentes, F.; Bhargava, A. Morphological analysis of quinoa germplasm grown under lowland desert conditions. *J. Agron. Crop Sci.* **2011**, *197*, 124–134. [CrossRef]
20. Gómez-Pando, L.R.; Álvarez-Castro, R.; Eguiluz-de la Barra, A. Short communication: Effect of salt stress on Peruvian germplasm of *Chenopodium quinoa* Willd.: A promising crop. *J. Agron. Crop Sci.* **2010**, *196*, 391–396. [CrossRef]
21. Ortiz, R.; Ruiz-Tapia, E.N.; Mujica-Sanchez, A. Sampling strategy for a core collection of Peruvian quinoa germplasm. *Theor. Appl. Genet.* **1998**, *96*, 475–483. [CrossRef] [PubMed]
22. Rodriguez, L.A.; Isla, M.T. Comparative analysis of genetic and morphologic diversity among quinoa accessions (*Chenopodium quinoa* Willd.) of the South of Chile and highland accessions. *J. Plant Breed. Crop Sci.* **2009**, *1*, 210–216.
23. Rojas, W. Multivariate analysis of genetic diversity of Bolivian quinoa germplasm. *Food Rev. Int.* **2003**, *19*, 9–23. [CrossRef]
24. Bazile, D.; Jacobsen, S.-E.; Verniau, A. The global expansion of quinoa: Trends and limits. *Front. Plant Sci.* **2016**, *7*, 622. [CrossRef] [PubMed]
25. Rojas, W.; Pinto, M. Ex situ conservation of quinoa: The Bolivian experience. In *Quinoa: Improvement and Sustainable Production*; Murphy, K., Matanguihan, J., Eds.; John Wiley & Sons, Inc.: Hoboken, NJ, USA, 2015; pp. 125–160, ISBN 978-1-118-62804-1.
26. Adolf, V.I.; Jacobsen, S.-E.; Shabala, S. Salt tolerance mechanisms in quinoa (*Chenopodium quinoa* Willd.). *Environ. Exp. Bot.* **2013**, *92*, 43–54. [CrossRef]
27. Bhargava, A.; Shukla, S.; Ohri, D. *Chenopodium quinoa*—An Indian perspective. *Ind. Crop. Prod.* **2006**, *23*, 73–87. [CrossRef]
28. Choukr-Allah, R.; Rao, N.K.; Hirich, A.; Shahid, M.; Alshankiti, A.; Toderich, K.; Gill, S.; Butt, K.U.R. Quinoa for marginal environments: Toward future food and nutritional security in MENA and central Asia regions. *Front. Plant Sci.* **2016**, *7*, 346. [CrossRef] [PubMed]
29. Jacobsen, S.E.; Mujica, A.; Jensen, C.R. The resistance of quinoa (*Chenopodium quinoa* Willd.) to adverse abiotic factors. *Food Rev. Int.* **2003**, *19*, 99–109. [CrossRef]
30. Ruiz, K.B.; Biondi, S.; Oses, R.; Acuña-Rodríguez, I.S.; Antognoni, F.; Martinez-Mosqueira, E.A.; Coulibaly, A.; Canahua-Murillo, A.; Pinto, M.; Zurita-Silva, A.; et al. Quinoa biodiversity and sustainability for food security under climate change. A review. *Agron. Sustain. Dev.* **2014**, *34*, 349–359. [CrossRef]
31. Trognitz, B.R. Prospects of breeding quinoa for tolerance to abiotic Stress. *Food Rev. Int.* **2003**, *19*, 129–137. [CrossRef]

32. Tuisima, L.L.C.; Fernández, C.E. An Andean ancient crop, *Chenopodium quinoa* Willd: A review. *Agric. Trop. Subtrop.* **2015**, *47*, 142–146. [CrossRef]
33. Ruiz, K.B.; Biondi, S.; Martínez, E.A.; Orsini, F.; Antognoni, F.; Jacobsen, S.-E. Quinoa—A model crop for understanding salt-tolerance mechanisms in halophytes. *Plant Biosyst. Int. J. Deal. All Asp. Plant Biol.* **2016**, *150*, 357–371. [CrossRef]
34. Jarvis, D.E.; Ho, Y.S.; Lightfoot, D.J.; Schmöckel, S.M.; Li, B.; Borm, T.J.A.; Ohyanagi, H.; Mineta, K.; Michell, C.T.; Saber, N.; et al. The genome of *Chenopodium quinoa*. *Nature* **2017**, *542*, 307–312. [CrossRef] [PubMed]
35. Blum, A. Drought resistance—Is it really a complex trait? *Funct. Plant Biol.* **2011**, *38*, 753–757. [CrossRef]
36. Aguilar, P.C.; Jacobsen, S.-E. Cultivation of quinoa on the Peruvian Altiplano. *Food Rev. Int.* **2003**, *19*, 31–41. [CrossRef]
37. Barrientos, E.; Carevic, F.; Delatorre, J. La sustentabilidad del altiplano sur de Bolivia y su relación con la ampliación de superficies de cultivo de quinua. *Idesia Arica* **2017**, *35*, 7–15. [CrossRef]
38. Geerts, S.; Raes, D.; Garcia, M.; Vacher, J.; Mamani, R.; Mendoza, J.; Huanca, R.; Morales, B.; Miranda, R.; Cusicanqui, J.; et al. Introducing deficit irrigation to stabilize yields of quinoa (*Chenopodium quinoa* Willd.). *Eur. J. Agron.* **2008**, *28*, 427–436. [CrossRef]
39. González, J.A.; Bruno, M.; Valoy, M.; Prado, F.E. Genotypic variation of gas exchange parameters and leaf stable carbon and nitrogen isotopes in ten quinoa cultivars grown under drought. *J. Agron. Crop Sci.* **2011**, *197*, 81–93. [CrossRef]
40. Martínez, E.A.; Veas, E.; Jorquera, C.; San Martín, R.; Jara, P. Re-Introduction of quinoa into Arid Chile: Cultivation of two lowland races under extremely low irrigation. *J. Agron. Crop Sci.* **2009**, *195*, 1–10. [CrossRef]
41. Vacher, J.-J. Responses of two main Andean crops, quinoa (*Chenopodium quinoa* Willd) and papa amarga (*Solanum juzepczukii* Buk.) to drought on the Bolivian Altiplano: Significance of local adaptation. *Agric. Ecosyst. Environ.* **1998**, *68*, 99–108. [CrossRef]
42. Bazile, D.; Pulvento, C.; Verniau, A.; Al-Nusairi, M.S.; Ba, D.; Breidy, J.; Hassan, L.; Mohammed, M.I.; Mambetov, O.; Otambekova, M.; et al. Worldwide evaluations of quinoa: Preliminary results from post International Year of Quinoa FAO projects in nine countries. *Front. Plant Sci.* **2016**, *7*, 850. [CrossRef] [PubMed]
43. Lavini, A.; Pulvento, C.; d'Andria, R.; Riccardi, M.; Choukr-Allah, R.; Belhabib, O.; Yazar, A.; İncekaya, ç.; Metin Sezen, S.; Qadir, M.; Jacobsen, S.-E. Quinoa's potential in the Mediterranean region. *J. Agron. Crop Sci.* **2014**, *200*, 344–360. [CrossRef]
44. Pulvento, C.; Riccardi, M.; Lavini, A.; Iafelice, G.; Marconi, E.; d'Andria, R. Yield and quality characteristics of quinoa grown in open field under different saline and non-saline irrigation regimes. *J. Agron. Crop Sci.* **2012**, *198*, 254–263. [CrossRef]
45. Pulvento, C.; Riccardi, M.; Lavini, A.; D'Andria, R.; Iafelice, G.; Marconi, E. Field trial evaluation of two *Chenopodium quinoa* genotypes grown under rain-fed conditions in a typical Mediterranean environment in south Italy. *J. Agron. Crop Sci.* **2010**, *196*, 407–411. [CrossRef]
46. Rambal, S.; Ratte, J.-P.; Mouillot, F.; Winkel, T. Trends in quinoa yield over the southern Bolivian Altiplano: Lessons from climate and land-use projections. In *Quinoa: Improvement and Sustainable Production*; Murphy, K., Matanguihan, J., Eds.; John Wiley & Sons, Inc.: Hoboken, NJ, USA, 2015; pp. 47–62, ISBN 978-1-118-62804-1.
47. Farooq, M.; Wahid, A.; Kobayashi, N.; Fujita, D.; Basra, S.M.A. Plant drought stress: Effects, mechanisms and management. *Agron. Sustain. Dev.* **2009**, *29*, 185–212. [CrossRef]
48. Alvarez-Flores, R.; Nguyen-Thi-Truc, A.; Peredo-Parada, S.; Joffre, R.; Winkel, T. Rooting plasticity in wild and cultivated Andean *Chenopodium* species under soil water deficit. *Plant Soil* **2018**, *425*, 479–492. [CrossRef]
49. Bascuñán-Godoy, L.; Reguera, M.; Abdel-Tawab, Y.M.; Blumwald, E. Water deficit stress-induced changes in carbon and nitrogen partitioning in *Chenopodium quinoa* Willd. *Planta* **2016**, *243*, 591–603. [CrossRef] [PubMed]
50. Bosque Sanchez, H.; Lemeur, R.; Damme, P.V.; Jacobsen, S.-E. Ecophysiological analysis of drought and salinity stress of quinoa (*Chenopodium quinoa* Willd.). *Food Rev. Int.* **2003**, *19*, 111–119. [CrossRef]
51. Cocozza, C.; Pulvento, C.; Lavini, A.; Riccardi, M.; d'Andria, R.; Tognetti, R. Effects of increasing salinity stress and decreasing water availability on ecophysiological traits of quinoa (*Chenopodium quinoa* Willd.) Grown in a mediterranean-type agroecosystem. *J. Agron. Crop Sci.* **2013**, *199*, 229–240. [CrossRef]

52. Jacobsen, S.-E.; Liu, F.; Jensen, C.R. Does root-sourced ABA play a role for regulation of stomata under drought in quinoa (*Chenopodium quinoa* Willd.). *Sci. Hortic.* **2009**, *122*, 281–287. [CrossRef]
53. Jensen, C.R.; Jacobsen, S.-E.; Andersen, M.N.; Nunez, N.; Andersen, S.D.; Rasmussen, L.; Mogensen, V.O. Leaf gas exchange and water relation characteristics of field quinoa (*Chenopodium quinoa* Willd.) during soil drying. *Eur. J. Agron.* **2000**, *13*, 11–25. [CrossRef]
54. Yang, A.; Akhtar, S.S.; Amjad, M.; Iqbal, S.; Jacobsen, S.-E. Growth and physiological responses of quinoa to drought and temperature stress. *J. Agron. Crop Sci.* **2016**, *202*, 445–453. [CrossRef]
55. Razzaghi, F.; Ahmadi, S.H.; Adolf, V.I.; Jensen, C.R.; Jacobsen, S.-E.; Andersen, M.N. Water relations and transpiration of quinoa (*Chenopodium quinoa* Willd.) under salinity and soil drying. *J. Agron. Crop Sci.* **2011**, *197*, 348–360. [CrossRef]
56. Sun, Y.; Liu, F.; Bendevis, M.; Shabala, S.; Jacobsen, S.-E. Sensitivity of two quinoa (*Chenopodium quinoa* Willd.) varieties to progressive drought stress. *J. Agron. Crop Sci.* **2014**, *200*, 12–23. [CrossRef]
57. Aguilar, P.C.; Cutipa, Z.; Machaca, E.; López, M.; Jacobsen, S.E. Variation of proline content of quinoa (*Chenopodium quinoa* Willd.) in high beds (Waru Waru). *Food Rev. Int.* **2003**, *19*, 121–127. [CrossRef]
58. González, J.A.; Gallardo, M.; Hilal, M.; Rosa, M.; Prado, F.E. Physiological responses of quinoa (*Chenopodium quinoa* Willd.) to drought and waterlogging stress: Dry matter partitioning. *Bot. Stud.* **2009**, *50*, 35–42.
59. Muscolo, A.; Panuccio, M.R.; Gioffrè, A.M.; Jacobsen, S.-E. Drought and salinity differently affect growth and secondary metabolites of "*Chenopodium quinoa* Willd" seedlings. In *Halophytes for Food Security in Dry Lands*; Khan, M.A., Ozturk, M., Gul, B., Ahmed, M.Z., Eds.; Elsevier Inc.: San Diego, CA, USA, 2016; pp. 259–275, ISBN 978-0-12-801854-5.
60. Killi, D.; Haworth, M. Diffusive and metabolic constraints to photosynthesis in quinoa during drought and salt stress. *Plants* **2017**, *6*, 49. [CrossRef] [PubMed]
61. Miranda-Apodaca, J.; Yoldi-Achalandabaso, A.; Aguirresarobe, A.; del Canto, A.; Pérez-López, U. Similarities and differences between the responses to osmotic and ionic stress in quinoa from a water use perspective. *Agric. Water Manag.* **2018**, *203*, 344–352. [CrossRef]
62. Geerts, S.; Raes, D.; Garcia, M.; Mendoza, J.; Huanca, R. Crop water use indicators to quantify the flexible phenology of quinoa (*Chenopodium quinoa* Willd.) in response to drought stress. *Field Crop. Res.* **2008**, *108*, 150–156. [CrossRef]
63. Reddy, A.R.; Chaitanya, K.V.; Vivekanandan, M. Drought-induced responses of photosynthesis and antioxidant metabolism in higher plants. *J. Plant Physiol.* **2004**, *161*, 1189–1202. [CrossRef]
64. Centritto, M.; Lauteri, M.; Monteverdi, M.C.; Serraj, R. Leaf gas exchange, carbon isotope discrimination, and grain yield in contrasting rice genotypes subjected to water deficits during the reproductive stage. *J. Exp. Bot.* **2009**, *60*, 2325–2339. [CrossRef] [PubMed]
65. Winkel, T.; Méthy, M.; Thénot, F. Radiation use efficiency, chlorophyll fluorescence, and reflectance Indices associated with ontogenic changes in water-limited *Chenopodium quinoa* leaves. *Photosynthetica* **2002**, *40*, 227–232. [CrossRef]
66. Fghire, R.; Anaya, F.; Ali, O.I.; Benlhabib, O.; Ragab, R.; Wahbi, S. Physiological and photosynthetic response of quinoa to drought stress. *Chil. J. Agric. Res.* **2015**, *75*, 174–183. [CrossRef]
67. Alvarez-Flores, R.; Winkel, T.; Degueldre, D.; Del Castillo, C.; Joffre, R. Plant growth dynamics and root morphology of little-known species of *Chenopodium* from contrasted Andean habitats. *Botany* **2013**, *92*, 101–108. [CrossRef]
68. Alvarez-Flores, R.; Winkel, T.; Nguyen-Thi-Truc, A.; Joffre, R. Root foraging capacity depends on root system architecture and ontogeny in seedlings of three Andean *Chenopodium* species. *Plant Soil* **2014**, *380*, 415–428. [CrossRef]
69. Razzaghi, F.; Plauborg, F.; Jacobsen, S.-E.; Jensen, C.R.; Andersen, M.N. Effect of nitrogen and water availability of three soil types on yield, radiation use efficiency and evapotranspiration in field-grown quinoa. *Agric. Water Manag.* **2012**, *109*, 20–29. [CrossRef]
70. Al-Naggar, A.M.; Abd El-Salam, R.M.; Badran, A.; El-Moghazi, M. Genotype and drought effects on morphological, physiological and yield traits of quinoa (*Chenopodium quinoa* Willd.). *Asian J. Adv. Agric. Res.* **2017**, *3*, 1–15. [CrossRef]
71. Walters, H.; Carpenter-Boggs, L.; Desta, K.; Yan, L.; Matanguihan, J.; Murphy, K. Effect of irrigation, intercrop, and cultivar on agronomic and nutritional characteristics of quinoa. *Agroecol. Sustain. Food Syst.* **2016**, *40*, 783–803. [CrossRef]

72. Piccinni, G.; Ko, J.; Marek, T.; Howell, T. Determination of growth-stage-specific crop coefficients (KC) of maize and sorghum. *Agric. Water Manag.* **2009**, *96*, 1698–1704. [CrossRef]
73. Garcia, M.; Raes, D.; Jacobsen, S.-E. Evapotranspiration analysis and irrigation requirements of quinoa (*Chenopodium quinoa*) in the Bolivian highlands. *Agric. Water Manag.* **2003**, *60*, 119–134. [CrossRef]
74. Barrios-Masias, F.H.; Jackson, L.E. Increasing the effective use of water in processing tomatoes through alternate furrow irrigation without a yield decrease. *Agric. Water Manag.* **2016**, *177*, 107–117. [CrossRef]
75. Geerts, S.; Raes, D.; Garcia, M.; Condori, O.; Mamani, J.; Miranda, R.; Cusicanqui, J.; Taboada, C.; Yucra, E.; Vacher, J. Could deficit irrigation be a sustainable practice for quinoa (*Chenopodium quinoa* Willd.) in the Southern Bolivian Altiplano? *Agric. Water Manag.* **2008**, *95*, 909–917. [CrossRef]
76. Alandia, G.; Jacobsen, S.-E.; Kyvsgaard, N.C.; Condori, B.; Liu, F. Nitrogen sustains seed yield of quinoa under intermediate drought. *J. Agron. Crop Sci.* **2016**, *202*, 281–291. [CrossRef]
77. Aziz, A.; Akram, N.A.; Ashraf, M. Influence of natural and synthetic vitamin C (ascorbic acid) on primary and secondary metabolites and associated metabolism in quinoa (*Chenopodium quinoa* Willd.) plants under water deficit regimes. *Plant Physiol. Biochem.* **2018**, *123*, 192–203. [CrossRef] [PubMed]
78. Hirich, A.; Choukr-Allah, R.; Jacobsen, S.-E. Deficit irrigation and organic compost improve growth and yield of quinoa and pea. *J. Agron. Crop Sci.* **2014**, *200*, 390–398. [CrossRef]
79. Hirich, A.; Choukr-Allah, R.; Jacobsen, S.-E. The combined effect of deficit irrigation by treated wastewater and organic amendment on quinoa (*Chenopodium quinoa* Willd.) productivity. *Desalination Water Treat.* **2014**, *52*, 2208–2213. [CrossRef]
80. Ramzani, P.M.A.; Shan, L.; Anjum, S.; Khan, W.-D.; Ronggui, H.; Iqbal, M.; Virk, Z.A.; Kausar, S. Improved quinoa growth, physiological response, and seed nutritional quality in three soils having different stresses by the application of acidified biochar and compost. *Plant Physiol. Biochem.* **2017**, *116*, 127–138. [CrossRef] [PubMed]
81. Kammann, C.I.; Linsel, S.; Gößling, J.W.; Koyro, H.-W. Influence of biochar on drought tolerance of *Chenopodium quinoa* Willd and on soil–plant relations. *Plant Soil* **2011**, *345*, 195–210. [CrossRef]
82. Akram, N.A.; Shafiq, F.; Ashraf, M. Ascorbic acid-A potential oxidant scavenger and its role in plant development and abiotic stress tolerance. *Front. Plant Sci.* **2017**, *8*, 613. [CrossRef] [PubMed]
83. Elewa, T.A.; Sadak, M.S.; Saad, A.M. Proline treatment improves physiological responses in quinoa plants under drought stress. *Biosci. Res.* **2017**, *14*, 21–33.
84. Iqbal, H.; Yaning, C.; Waqas, M.; Rehman, H.; Shareef, M.; Iqbal, S. Hydrogen peroxide application improves quinoa performance by affecting physiological and biochemical mechanisms under water-deficit conditions. *J. Agron. Crop Sci.* **2018**, 1–13. [CrossRef]
85. Verma, G.; Mishra, S.; Sangwan, N.; Sharma, S. Reactive oxygen species mediate axis-cotyledon signaling to induce reserve mobilization during germination and seedling establishment in *Vigna radiata*. *J. Plant Physiol.* **2015**, *184*, 79–88. [CrossRef] [PubMed]
86. González-Teuber, M.; Urzúa, A.; Plaza, P.; Bascuñán-Godoy, L. Effects of root endophytic fungi on response of *Chenopodium quinoa* to drought stress. *Plant Ecol.* **2018**, *219*, 231–240. [CrossRef]
87. Hussin, S.; Khalifa, W.; Geissler, N.; Koyro, H.-W. Influence of the root endophyte *Piriformospora indica* on the plant water relations, gas exchange and growth of *Chenopodium quinoa* at limited water availability. *J. Agron. Crop Sci.* **2017**, *203*, 373–384. [CrossRef]
88. Pitzschke, A. Developmental peculiarities and seed-borne Endophytes in Quinoa: Omnipresent, robust bacilli contribute to plant fitness. *Front. Microbiol.* **2016**, *7*. [CrossRef] [PubMed]
89. Pitzschke, A. Molecular dynamics in germinating, endophyte-colonized quinoa seeds. *Plant Soil* **2018**, *422*, 135–154. [CrossRef] [PubMed]
90. González-Teuber, M.; Vilo, C.; Bascuñán-Godoy, L. Molecular characterization of endophytic fungi associated with the roots of *Chenopodium quinoa* inhabiting the Atacama Desert, Chile. *Genom. Data* **2017**, *11*, 109–112. [CrossRef] [PubMed]
91. González, J.A.; Konishi, Y.; Bruno, M.; Valoy, M.; Prado, F.E. Interrelationships among seed yield, total protein and amino acid composition of ten quinoa (*Chenopodium quinoa*) cultivars from two different agroecological regions. *J. Sci. Food Agric.* **2012**, *92*, 1222–1229. [CrossRef] [PubMed]
92. Prado, F.E.; Fernández-Turiel, J.L.; Tsarouchi, M.; Psaras, G.K.; González, J.A. Variation of seed mineral concentrations in seven quinoa cultivars grown in two agroecological sites. *Cereal Chem.* **2014**, *91*, 453–459. [CrossRef]

93. Reguera, M.; Conesa, C.M.; Gómez, A.G.; Haros, C.M.; Casas, M.Á.P.; Labarca, V.B.; Rosa, L.B.; Mangas, I.B.; Mujica, Á.; Godoy, L.B. The impact of different agroecological conditions on the nutritional composition of quinoa seeds. *PeerJ Prepr.* **2018**, *6*. [CrossRef] [PubMed]
94. Fischer, S.; Wilckens, R.; Jara, J.; Aranda, M. Variation in antioxidant capacity of quinoa (*Chenopodium quinoa* Will) subjected to drought stress. *Ind. Crop. Prod.* **2013**, *46*, 341–349. [CrossRef]
95. Fischer, S.; Wilckens, R.; Jara, J.; Aranda, M.; Valdivia, W.; Bustamante, L.; Graf, F.; Obal, I. Protein and antioxidant composition of quinoa (*Chenopodium quinoa* Willd.) sprout from seeds submitted to water stress, salinity and light conditions. *Ind. Crop. Prod.* **2017**, *107*, 558–564. [CrossRef]
96. Raney, J.A.; Reynolds, D.J.; Elzinga, D.B.; Page, J.; Udall, J.A.; Jellen, E.N.; Bonfacio, A.; Fairbanks, D.J.; Maughan, P.J. Transcriptome analysis of drought induced stress in *Chenopodium quinoa*. *Am. J. Plant Sci.* **2014**, *5*, 338–357. [CrossRef]
97. Davin, L.B.; Lewis, N.G. Dirigent proteins and dirigent sites explain the mystery of specificity of radical precursor coupling in lignan and lignin biosynthesis. *Plant Physiol.* **2000**, *123*, 453–462. [CrossRef] [PubMed]
98. Vierling, E. The roles of heat shock proteins in plants. *Annu. Rev. Plant Physiol. Plant Mol. Biol.* **1991**, *42*, 579–620. [CrossRef]
99. Wang, W.; Vinocur, B.; Shoseyov, O.; Altman, A. Role of plant heat-shock proteins and molecular chaperones in the abiotic stress response. *Trends Plant Sci.* **2004**, *9*, 244–252. [CrossRef] [PubMed]
100. Cho, E.; Choi, Y. A nuclear-localized HSP70 confers thermoprotective activity and drought-stress tolerance on plants. *Biotechnol. Lett.* **2009**, *31*, 597–606. [CrossRef] [PubMed]
101. Liu, J.; Wang, R.; Liu, W.; Zhang, H.; Guo, Y.; Wen, R. Genome-wide characterization of heat-shock protein 70s from *Chenopodium quinoa* and expression analyses of Cqhsp70s in response to drought stress. *Genes* **2018**, *9*. [CrossRef] [PubMed]
102. Morales, A.; Zurita-Silva, A.; Maldonado, J.; Silva, H. Transcriptional responses of Chilean quinoa (*Chenopodium quinoa* Willd.) under water deficit conditions uncovers ABA-independent expression patterns. *Front. Plant Sci.* **2017**, *8*, 216. [CrossRef] [PubMed]
103. Rhoades, J.D. Salinity: Electrical conductivity and total dissolved solids. In *Methods of Soil Analysis. Part 3. Chemical Methods*; Sparks, D.L., Ed.; American Society of Agronomy, Crop Science Society of America, Soil Science Society of America: Madison, WI, USA, 1996; pp. 417–435.
104. LiLi, L.; QiYan, J.; FengJuan, N.; Zheng, H.; Hui, Z. Research progress on salt tolerance mechanisms in quinoa. *J. Agric. Sci. Technol. Beijing* **2016**, *18*, 31–40.
105. Gupta, B.; Huang, B. Mechanism of salinity tolerance in plants: Physiological, biochemical, and molecular characterization. *Int. J. Genom.* **2014**, *18*. [CrossRef] [PubMed]
106. Munns, R. Comparative physiology of salt and water stress. *Plant Cell Environ.* **2002**, *25*, 239–250. [CrossRef] [PubMed]
107. Munns, R.; Gilliham, M. Salinity tolerance of crops—What is the cost? *New Phytol.* **2015**, *208*, 668–673. [CrossRef] [PubMed]
108. Miller, G.; Suzuki, N.; Ciftci-Yilmaz, S.; Mittler, R. Reactive oxygen species homeostasis and signalling during drought and salinity stresses. *Plant Cell Environ.* **2010**, *33*, 453–467. [CrossRef] [PubMed]
109. Mittler, R. Oxidative stress, antioxidants and stress tolerance. *Trends Plant Sci.* **2002**, *7*, 405–410. [CrossRef]
110. Choudhury, F.K.; Rivero, R.M.; Blumwald, E.; Mittler, R. Reactive oxygen species, abiotic stress and stress combination. *Plant J.* **2017**, *90*, 856–867. [CrossRef] [PubMed]
111. Sharma, P.; Jha, A.B.; Dubey, R.S.; Pessarakli, M. Reactive oxygen species, oxidative damage, and antioxidative defense mechanism in plants under stressful conditions. *J. Bot.* **2012**, *2012*, 1–27. [CrossRef]
112. Gunes, A.; Inal, A.; Alpaslan, M.; Eraslan, F.; Bagci, E.G.; Cicek, N. Salicylic acid induced changes on some physiological parameters symptomatic for oxidative stress and mineral nutrition in maize (*Zea mays* L.) grown under salinity. *J. Plant Physiol.* **2007**, *164*, 728–736. [CrossRef] [PubMed]
113. Peterson, A.; Murphy, K. Tolerance of lowland quinoa cultivars to sodium chloride and sodium sulfate salinity. *Crop Sci.* **2015**, *55*, 331–338. [CrossRef]
114. Sairam, R.K.; Rao, K.V.; Srivastava, G.C. Differential response of wheat genotypes to long term salinity stress in relation to oxidative stress, antioxidant activity and osmolyte concentration. *Plant Sci.* **2002**, *163*, 1037–1046. [CrossRef]
115. Shannon, M.C.; Grieve, C.M. Tolerance of vegetable crops to salinity. *Sci. Hortic.* **1998**, *78*, 5–38. [CrossRef]

116. Aloisi, I.; Parrotta, L.; Ruiz, K.B.; Landi, C.; Bini, L.; Cai, G.; Biondi, S.; Del Duca, S. New insight into quinoa seed quality under salinity: Changes in proteomic and amino acid profiles, phenolic content, and antioxidant activity of protein extracts. *Front. Plant Sci.* **2016**, *7*, 656. [CrossRef] [PubMed]
117. González, J.A.; Prado, F.E. Germination in relation to temperature and different salt concentration in *Chenopodium quinoa*. *Agrochimica* **1992**, *36*, 101–107.
118. Schmöckel, S.M.; Lightfoot, D.J.; Razali, R.; Tester, M.; Jarvis, D.E. Identification of putative transmembrane proteins involved in salinity tolerance in *Chenopodium quinoa* by integrating physiological data, RNAseq, and SNP analyses. *Front. Plant Sci.* **2017**, *8*. [CrossRef] [PubMed]
119. Shabala, S.; Hariadi, Y.; Jacobsen, S.-E. Genotypic difference in salinity tolerance in quinoa is determined by differential control of xylem Na+ loading and stomatal density. *J. Plant Physiol.* **2013**, *170*, 906–914. [CrossRef] [PubMed]
120. Orsini, F.; Accorsi, M.; Gianquinto, G.; Dinelli, G.; Antognoni, F.; Carrasco, K.B.R.; Martinez, E.A.; Alnayef, M.; Marotti, I.; Bosi, S.; et al. Beyond the ionic and osmotic response to salinity in *Chenopodium quinoa*: Functional elements of successful halophytism. *Funct. Plant Biol.* **2011**, *38*, 818–831. [CrossRef]
121. Ruiz, K.B.; Aloisi, I.; Del Duca, S.; Canelo, V.; Torrigiani, P.; Silva, H.; Biondi, S. Salares versus coastal ecotypes of quinoa: Salinity responses in Chilean landraces from contrasting habitats. *Plant Physiol. Biochem.* **2016**, *101*, 1–13. [CrossRef] [PubMed]
122. Ruiz, K.B.; Rapparini, F.; Bertazza, G.; Silva, H.; Torrigiani, P.; Biondi, S. Comparing salt-induced responses at the transcript level in a salares and coastal-lowlands landrace of quinoa (*Chenopodium quinoa* Willd). *Environ. Exp. Bot.* **2017**, *139*, 127–142. [CrossRef]
123. Sun, Y.; Lindberg, S.; Shabala, L.; Morgan, S.; Shabala, S.; Jacobsen, S.-E. A comparative analysis of cytosolic Na+ changes under salinity between halophyte quinoa (*Chenopodium quinoa*) and glycophyte pea (*Pisum sativum*). *Environ. Exp. Bot.* **2017**, *141*, 154–160. [CrossRef]
124. Eisa, S.S.; Eid, M.A.; Abd, E.-S.; Hussin, S.A.; Abdel-Ati, A.A.; El-Bordeny, N.E.; Ali, S.H.; Al-Sayed, H.M.A.; Lotfy, M.E.; Masoud, A.M.; et al. *Chenopodium quinoa* Willd. A new cash crop halophyte for saline regions of Egypt. *Aust. J. Crop Sci.* **2017**, *11*, 343–351. [CrossRef]
125. Hariadi, Y.; Marandon, K.; Tian, Y.; Jacobsen, S.-E.; Shabala, S. Ionic and osmotic relations in quinoa (*Chenopodium quinoa* Willd.) plants grown at various salinity levels. *J. Exp. Bot.* **2011**, *62*, 185–193. [CrossRef] [PubMed]
126. Debez, A.; Hamed, K.B.; Grignon, C.; Abdelly, C. Salinity effects on germination, growth, and seed production of the halophyte *Cakile maritima*. *Plant Soil* **2004**, *262*, 179–189. [CrossRef]
127. Gul, B.; Ansari, R.; Flowers, T.J.; Khan, M.A. Germination strategies of halophyte seeds under salinity. *Environ. Exp. Bot.* **2013**, *92*, 4–18. [CrossRef]
128. Delatorre-Herrera, J.; Pinto, M. Importance of ionic and osmotic components of salt stress on the germination of four quinoa (*Chenopodium quinoa* Willd.) selections. *Chil. J. Agric. Res.* **2009**, *69*, 477–485. [CrossRef]
129. Panuccio, M.R.; Jacobsen, S.E.; Akhtar, S.S.; Muscolo, A. Effect of saline water on seed germination and early seedling growth of the halophyte quinoa. *AoB PLANTS* **2014**, *6*. [CrossRef] [PubMed]
130. Prado, F.E.; Boero, C.; Gallardo, M.; González, J.A. Effect of NaCl on germination, growth, and soluble sugar content in *Chenopodium quinoa* Willd. seeds. *Bot. Bull. Acad. Sin.* **2000**, *41*, 27–34.
131. Ruffino, A.M.C.; Rosa, M.; Hilal, M.; González, J.A.; Prado, F.E. The role of cotyledon metabolism in the establishment of quinoa (*Chenopodium quinoa*) seedlings growing under salinity. *Plant Soil* **2010**, *326*, 213–224. [CrossRef]
132. Ruiz-Carrasco, K.; Antognoni, F.; Coulibaly, A.K.; Lizardi, S.; Covarrubias, A.; Martínez, E.A.; Molina-Montenegro, M.A.; Biondi, S.; Zurita-Silva, A. Variation in salinity tolerance of four lowland genotypes of quinoa (*Chenopodium quinoa* Willd.) as assessed by growth, physiological traits, and sodium transporter gene expression. *Plant Physiol. Biochem.* **2011**, *49*, 1333–1341. [CrossRef] [PubMed]
133. Rosa, M.; Hilal, M.; González, J.A.; Prado, F.E. Changes in soluble carbohydrates and related enzymes induced by low temperature during early developmental stages of quinoa (*Chenopodium quinoa*) seedlings. *J. Plant Physiol.* **2004**, *161*, 683–689. [CrossRef] [PubMed]
134. Dinneny, J.R. Traversing organizational scales in plant salt-stress responses. *Curr. Opin. Plant Biol.* **2015**, *23*, 70–75. [CrossRef] [PubMed]
135. Adolf, V.I.; Shabala, S.; Andersen, M.N.; Razzaghi, F.; Jacobsen, S.-E. Varietal differences of quinoa's tolerance to saline conditions. *Plant Soil* **2012**, *357*, 117–129. [CrossRef]

136. Becker, V.I.; Goessling, J.W.; Duarte, B.; Caçador, I.; Liu, F.; Rosenqvist, E.; Jacobsen, S.-E. Combined effects of soil salinity and high temperature on photosynthesis and growth of quinoa plants (*Chenopodium quinoa*). *Funct. Plant Biol.* **2017**, *44*, 665–678. [CrossRef]
137. Eisa, S.; Hussin, S.; Geissler, N.; Koyro, H.W. Effect of NaCl salinity on water relations, photosynthesis and chemical composition of Quinoa (*Chenopodium quinoa* Willd.) as a potential cash crop halophyte. *Aust. J. Crop Sci.* **2012**, *6*, 357–368.
138. Talebnejad, R.; Sepaskhah, A.R. Physiological characteristics, gas exchange, and plant ion relations of quinoa to different saline groundwater depths and water salinity. *Arch. Agron. Soil Sci.* **2016**, *62*, 1347–1367. [CrossRef]
139. Geissler, N.; Hussin, S.; El-Far, M.M.M.; Koyro, H.-W. Elevated atmospheric CO2 concentration leads to different salt resistance mechanisms in a C3 (*Chenopodium quinoa*) and a C4 (*Atriplex nummularia*) halophyte. *Environ. Exp. Bot.* **2015**, *118*, 67–77. [CrossRef]
140. Yang, A.; Akhtar, S.S.; Iqbal, S.; Amjad, M.; Naveed, M.; Zahir, Z.A.; Jacobsen, S.-E. Enhancing salt tolerance in quinoa by halotolerant bacterial inoculation. *Funct. Plant Biol.* **2016**, *43*, 632–642. [CrossRef]
141. Yang, A.; Akhtar, S.S.; Iqbal, S.; Qi, Z.; Alandia, G.; Saddiq, M.S.; Jacobsen, S.-E. Saponin seed priming improves salt tolerance in quinoa. *J. Agron. Crop Sci.* **2018**, *204*, 31–39. [CrossRef]
142. Li, H.; Lei, P.; Pang, X.; Li, S.; Xu, H.; Xu, Z.; Feng, X. Enhanced tolerance to salt stress in canola (*Brassica napus* L.) seedlings inoculated with the halotolerant Enterobacter cloacae HSNJ4. *Appl. Soil Ecol.* **2017**, *119*, 26–34. [CrossRef]
143. Ashraf, M.; Foolad, M.R. Pre-sowing seed treatment—A shotgun approach to improve germination, plant growth, and crop yield under saline and non-saline conditions. In *Advances in Agronomy*; Academic Press: Waltham, MA, USA, 2005; Volume 88, pp. 223–271.
144. Moreno, C.; Seal, C.E.; Papenbrock, J. Seed priming improves germination in saline conditions for *Chenopodium quinoa* and *Amaranthus caudatus*. *J. Agron. Crop Sci.* **2018**, *204*, 40–48. [CrossRef]
145. Gómez, M.B.; Castro, P.A.; Mignone, C.; Bertero, H.D. Can yield potential be increased by manipulation of reproductive partitioning in quinoa (*Chenopodium quinoa*)? Evidence from gibberellic acid synthesis inhibition using Paclobutrazol. *Funct. Plant Biol.* **2011**, *38*, 420. [CrossRef]
146. Waqas, M.; Yaning, C.; Iqbal, H.; Shareef, M.; Rehman, H.; Yang, Y. Paclobutrazol improves salt tolerance in quinoa: Beyond the stomatal and biochemical interventions. *J. Agron. Crop Sci.* **2017**, *203*, 315–322. [CrossRef]
147. Kiani-Pouya, A.; Roessner, U.; Jayasinghe, N.S.; Lutz, A.; Rupasinghe, T.; Bazihizina, N.; Bohm, J.; Alharbi, S.; Hedrich, R.; Shabala, S. Epidermal bladder cells confer salinity stress tolerance in the halophyte quinoa and *Atriplex species*. *Plant Cell Environ.* **2017**, *40*, 1900–1915. [CrossRef] [PubMed]
148. Zou, C.; Chen, A.; Xiao, L.; Muller, H.M.; Ache, P.; Haberer, G.; Zhang, M.; Jia, W.; Deng, P.; Huang, R.; et al. A high-quality genome assembly of quinoa provides insights into the molecular basis of salt bladder-based salinity tolerance and the exceptional nutritional value. *Cell Res.* **2017**, *27*, 1327–1340. [CrossRef] [PubMed]
149. Koyro, H.-W.; Eisa, S.S. Effect of salinity on composition, viability and germination of seeds of *Chenopodium quinoa* Willd. *Plant Soil* **2008**, *302*, 79–90. [CrossRef]
150. Shabala, S.; Bose, J.; Hedrich, R. Salt bladders: Do they matter? *Trends Plant Sci.* **2014**, *19*, 687–691. [CrossRef] [PubMed]
151. Shabala, L.; Mackay, A.; Tian, Y.; Jacobsen, S.-E.; Zhou, D.; Shabala, S. Oxidative stress protection and stomatal patterning as components of salinity tolerance mechanism in quinoa (*Chenopodium quinoa*). *Physiol. Plant.* **2012**, *146*, 26–38. [CrossRef] [PubMed]
152. Jou, Y.; Wang, Y.-L.; Yen, H.E. Vacuolar acidity, protein profile, and crystal composition of epidermal bladder cells of the halophyte *Mesembryanthemum crystallinum*. *Funct. Plant Biol.* **2007**, *34*, 353–359. [CrossRef]
153. Barkla, B.J.; Vera-Estrella, R.; Pantoja, O. Protein profiling of epidermal bladder cells from the halophyte *Mesembryanthemum crystallinum*. *Preotomics* **2012**, *12*, 2862–2865. [CrossRef] [PubMed]
154. Freitas, H.; Breckle, S.-W. Importance of bladder hairs for salt tolerance of field-grown *Atriplex species* from a Portuguese salt marsh. *Flora* **1992**, *187*, 283–297. [CrossRef]
155. LoPresti, E.F. Chenopod salt bladders deter insect herbivores. *Oecologia* **2014**, *174*, 921–930. [CrossRef] [PubMed]
156. Bonales-Alatorre, E.; Shabala, S.; Chen, Z.-H.; Pottosin, I. Reduced tonoplast fast-activating and slow-activating channel activity is essential for conferring salinity tolerance in a facultative halophyte, quinoa. *Plant Physiol.* **2013**, *162*, 940–952. [CrossRef] [PubMed]

157. Bonales-Alatorre, E.; Pottosin, I.; Shabala, L.; Chen, Z.-H.; Zeng, F.; Jacobsen, S.-E.; Shabala, S. Differential activity of plasma and vacuolar membrane transporters contributes to genotypic differences in salinity tolerance in a halophyte species, *Chenopodium quinoa*. *Int. J. Mol. Sci.* **2013**, *14*, 9267–9285. [CrossRef] [PubMed]
158. Iqbal, S.; Basra, S.M.A.; Afzal, I.; Wahid, A. Exploring potential of well adapted quinoa lines for salt tolerance. *Int. J. Agric. Biol.* **2017**, *19*, 933–940. [CrossRef]
159. Saleem, M.A.; Basra, S.M.A.; Afzal, I.; Iqbal, S.; Sohail, S.; Naz, S. Exploring the potential of quinoa accessions for salt tolerance in soilless culture. *Int. J. Agric. Biol.* **2017**, *19*, 233–240. [CrossRef]
160. Ismail, H.; Maksimović, J.D.; Maksimović, V.; Shabala, L.; Živanović, B.D.; Tian, Y.; Jacobsen, S.-E.; Shabala, S. Rutin, a flavonoid with antioxidant activity, improves plant salinity tolerance by regulating K+ retention and Na+ exclusion from leaf mesophyll in quinoa and broad beans. *Funct. Plant Biol.* **2016**, *43*, 75–86. [CrossRef]
161. Pottosin, I.; Bonales-Alatorre, E.; Shabala, S. Choline but not its derivative betaine blocks slow vacuolar channels in the halophyte *Chenopodium quinoa*: Implications for salinity stress responses. *FEBS Lett.* **2014**, *588*, 3918–3923. [CrossRef] [PubMed]
162. Zhu, J.-K. Regulation of ion homeostasis under salt stress. *Curr. Opin. Plant Biol.* **2003**, *6*, 441–445. [CrossRef]
163. Gómez-Caravaca, A.M.; Iafelice, G.; Lavini, A.; Pulvento, C.; Caboni, M.F.; Marconi, E. Phenolic compounds and saponins in quinoa samples (*Chenopodium quinoa* Willd.) grown under different saline and nonsaline irrigation regimens. *J. Agric. Food Chem.* **2012**, *60*, 4620–4627. [CrossRef] [PubMed]
164. Karyotis, T.; Iliadis, C.; Noulas, C.; Mitsibonas, T. Preliminary research on seed production and nutrient content for certain quinoa varieties in a saline–sodic soil. *J. Agron. Crop Sci.* **2003**, *189*, 402–408. [CrossRef]
165. Wu, G.; Peterson, A.J.; Morris, C.F.; Murphy, K.M. Quinoa seed quality response to sodium chloride and sodium sulfate salinity. *Front. Plant Sci.* **2016**, *7*. [CrossRef] [PubMed]
166. Yasui, Y.; Hirakawa, H.; Oikawa, T.; Toyoshima, M.; Matsuzaki, C.; Ueno, M.; Mizuno, N.; Nagatoshi, Y.; Imamura, T.; Miyago, M.; et al. Draft genome sequence of an inbred line of *Chenopodium quinoa*, an allotetraploid crop with great environmental adaptability and outstanding nutritional properties. *DNA Res.* **2016**, *23*, 535–546. [CrossRef] [PubMed]
167. Qiu, Q.-S.; Guo, Y.; Dietrich, M.A.; Schumaker, K.S.; Zhu, J.-K. Regulation of SOS1, a plasma membrane Na+/H+ exchanger in *Arabidopsis thaliana*, by SOS2 and SOS3. *Proc. Natl. Acad. Sci. USA* **2002**, *99*, 8436–8441. [CrossRef] [PubMed]
168. Shi, H.; Ishitani, M.; Kim, C.; Zhu, J.-K. The *Arabidopsis thaliana* salt tolerance gene SOS1 encodes a putative Na+/H+ antiporter. *Proc. Natl. Acad. Sci. USA* **2000**, *97*, 6896–6901. [CrossRef] [PubMed]
169. Shi, H.; Quintero, F.J.; Pardo, J.M.; Zhu, J.-K. The putative plasma membrane Na+/H+ antiporter SOS1 controls long-distance Na+ transport in plants. *Plant Cell* **2002**, *14*, 465–477. [CrossRef] [PubMed]
170. Apse, M.P.; Aharon, G.S.; Snedden, W.A.; Blumwald, E. Salt tolerance conferred by overexpression of a vacuolar Na+/H+ antiport in *Arabidopsis*. *Science* **1999**, *285*, 1256–1258. [CrossRef] [PubMed]
171. Blumwald, E.; Poole, R.J. Na+/H+ Antiport in isolated tonoplast vesicles from storage tissue of *Beta vulgaris*. *Plant Physiol.* **1985**, *78*, 163–167. [CrossRef] [PubMed]
172. Maughan, P.J.; Turner, T.B.; Coleman, C.E.; Elzinga, D.B.; Jellen, E.N.; Morales, J.A.; Udall, J.A.; Fairbanks, D.J.; Bonifacio, A. Characterization of Salt Overly Sensitive 1 (SOS1) gene homoeologs in quinoa (*Chenopodium quinoa* Willd.). *Genome* **2009**, *52*, 647–657. [CrossRef] [PubMed]
173. Morales, A.J.; Bajgain, P.; Garver, Z.; Maughan, P.J.; Udall, J.A. Physiological responses of *Chenopodium quinoa* to salt stress. *Int. J. Plant Physiol. Biochem.* **2011**, *3*, 219–232.
174. Jain, G.; Schwinn, K.E.; Gould, K.S. Betalain induction by l-DOPA application confers photoprotection to saline-exposed leaves of Disphyma australe. *New Phytol.* **2015**, *207*, 1075–1083. [CrossRef] [PubMed]
175. Imamura, T.; Takagi, H.; Miyazato, A.; Ohki, S.; Mizukoshi, H.; Mori, M. Isolation and characterization of the betalain biosynthesis gene involved in hypocotyl pigmentation of the allotetraploid *Chenopodium quinoa*. *Biochem. Biophys. Res. Commun.* **2018**, *5*, 280–286. [CrossRef] [PubMed]
176. Awasthi, R.; Bhandari, K.; Nayyar, H. Temperature stress and redox homeostasis in agricultural crops. *Front. Environ. Sci.* **2015**, *3*, 11. [CrossRef]
177. Sehgal, A.; Sita, K.; Kumar, J.; Kumar, S.; Singh, S.; Siddique, K.H.M.; Nayyar, H. Effects of drought, heat and their interaction on the growth, yield and photosynthetic function of lentil (*Lens culinaris* Medikus) genotypes varying in heat and drought sensitivity. *Front. Plant Sci.* **2017**, *8*, 1776. [CrossRef] [PubMed]

178. Wahid, A.; Gelani, S.; Ashraf, M.; Foolad, M.R. Heat tolerance in plants: An overview. *Environ. Exp. Bot.* **2007**, *61*, 199–223. [CrossRef]
179. Driedonks, N.; Rieu, I.; Vriezen, W.H. Breeding for plant heat tolerance at vegetative and reproductive stages. *Plant Reprod.* **2016**, *29*, 67–79. [CrossRef] [PubMed]
180. Prasad, V.; Bheemanahalli, R.; Jagadish, S.V.K. Field crops and the fear of heat stress—Opportunities, challenges and future directions. *Field Crop. Res.* **2017**, *200*, 114–121. [CrossRef]
181. Singh, R.P.; Prasad, P.V.V.; Sunita, K.; Giri, S.N.; Reddy, K.R. Influence of high temperature and breeding for heat tolerance in cotton: A review. *Adv. Agron.* **2007**, *93*, 313–385. [CrossRef]
182. Bita, C.E.; Gerats, T. Plant tolerance to high temperature in a changing environment: Scientific fundamentals and production of heat stress-tolerant crops. *Front. Plant Sci.* **2013**, *4*, 273. [CrossRef] [PubMed]
183. Kotak, S.; Larkindale, J.; Lee, U.; von Koskull-Döring, P.; Vierling, E.; Scharf, K.-D. Complexity of the heat stress response in plants. *Curr. Opin. Plant Biol.* **2007**, *10*, 310–316. [CrossRef] [PubMed]
184. Ohama, N.; Sato, H.; Shinozaki, K.; Yamaguchi-Shinozaki, K. Transcriptional regulatory network of plant heat stress response. *Trends Plant Sci.* **2017**, *22*, 53–65. [CrossRef] [PubMed]
185. Jacobsen, S.-E.; Monteros, C.; Christiansen, J.L.; Bravo, L.A.; Corcuera, L.J.; Mujica, A. Plant responses of quinoa (*Chenopodium quinoa* Willd.) to frost at various phenological stages. *Eur. J. Agron.* **2005**, *22*, 131–139. [CrossRef]
186. Isobe, K.; Uziie, K.; Hitomi, S.; Furuya, U. Agronomic studies on quinoa (*Chenopodium quinoa* Willd.) cultivation in Japan. *Jpn. J. Crop Sci.* **2012**, *81*, 167–172. [CrossRef]
187. Jacobsen, S.-E. The scope for adaptation of quinoa in Northern Latitudes of Europe. *J. Agron. Crop Sci.* **2017**, *203*, 603–613. [CrossRef]
188. Maliro, M.F.A.; Guwela, V.F.; Nyaika, J.; Murphy, K.M. Preliminary studies of the performance of quinoa (*Chenopodium quinoa* Willd.) genotypes under irrigated and rainfed conditions of central Malawi. *Front. Plant Sci.* **2017**, *8*, 227. [CrossRef] [PubMed]
189. Mosyakin, S.L.; Schwartau, V. Quinoa as a promising pseudocereal crop for Ukraine. *Agric. Sci. Pract.* **2015**, *2*, 3–11. [CrossRef]
190. Murphy, K.M.; Bazile, D.; Kellogg, J.; Rahmanian, M. Development of a worldwide consortium on evolutionary participatory breeding in quinoa. *Front. Plant Sci.* **2016**, *7*, 608. [CrossRef] [PubMed]
191. Spehar, C.R.; Santos, R.L. de B. Agronomic performance of quinoa selected in the Brazilian Savannah. *Pesqui. Agropecu. Bras.* **2005**, *40*, 609–612. [CrossRef]
192. Yazar, A.; Sezen, S.M.; Tekin, S.; Incekaya, C. Quinoa: From experimentation to production in Turkey. In *Quinoa for Future Food and Nutrition Security in Marginal Environments*; International Center for Biosaline Agriculture: Dubai, United Arab Emirates, 2016; p. 11.
193. Hirich, A.; Choukr-Allah, R.; Jacobsen, S.-E. Quinoa in Morocco—Effect of sowing dates on development and yield. *J. Agron. Crop Sci.* **2014**, *200*, 371–377. [CrossRef]
194. Präger, A.; Munz, S.; Nkebiwe, P.; Mast, B.; Graeff-Hönninger, S. Yield and quality characteristics of different quinoa (*Chenopodium quinoa* Willd.) Cultivars grown under field conditions in Southwestern Germany. *Agronomy* **2018**, *8*, 197. [CrossRef]
195. Pires, J.L. Avaliação do Comportamento Agronómico da Quinoa (*Chenopodium quinoa* Willd.), em Diferentes Regimes Hídricos e Níveis de Fertilização Azotada, nas Condições Agroecológicas de Trás-os-Montes. Master's Thesis, Instituto Politecnico de Bracanca Escola Superior Agraria, Bragança, Portugal, 2017.
196. Peterson, A.; Murphy, K.M. Quinoa cultivation for temperate North America: Considerations and areas for investigation. In *Quinoa: Improvement and Sustainable Production*; Murphy, K., Matanguihan, J., Eds.; John Wiley & Sons, Inc.: Hoboken, NJ, USA, 2015; pp. 173–192, ISBN 978-1-118-62804-1.
197. Chilo, G.; Vacca, M.; Carabajal, R.; Ochoa, M. Temperature and salinity effects on germination and seedling growth on two varieties of *Chenopodium quinoa*. *AgriScientia* **2009**, *26*, 15–22.
198. Coulibaly, A.; Sangare, A.; Konate, M.; Traore, S.; Ruiz, K.B.; Martinez, E.A. Assessment and adaptation of quinoa (*Chenopodium quinoa* Willd) to the agroclimatic conditions in Mali, West Africa: An example of South-North-South cooperation. In *State of the Art Report on Quinoa around the World 2013*; Bazile, D., Bertero, H.D., Nieto, C., Eds.; FAO: Santiago, Chile; CIRAD: Montpellier, France, 2015; Volume 1, pp. 524–533.
199. González, J.A.; Buedo, S.E.; Bruno, M.; Prado, F.E. Quantifying cardinal temperature in Quinoa (*Chenopodium quinoa*) cultivars. *Lilloa* **2017**, *54*, 179–194. [CrossRef]

200. Isobe, K.; Ishihara, M.; Nishigai, Y.; Miyagawa, N.; Higo, M.; Torigoe, Y. Effects of soil moisture, temperature and sowing depth on emergence of Quinoa (*Chenopodium quinoa* Willd.). *Jpn. J. Crop Sci.* **2015**, *84*, 17–21. [CrossRef]
201. Jacobsen, S.-E.; Jørnsgård, B.; Christiansen, J.L.; Stølen, O. Effect of harvest time, drying technique, temperature and light on the germination of quinoa (*Chenopodium quinoa*). *Seed Sci. Technol.* **1999**, *27*, 937–944.
202. Jacobsen, S.-E.; Bach, A.P. The influence of temperature on seed germination rate in quinoa (*Chenopodium quinoa* Willd.). *Seed Sci. Technol.* **1998**, *26*, 515–523.
203. Mamedi, A.; Tavakkol Afshari, R.; Oveisi, M. Cardinal temperatures for seed germination of three quinoa (*Chenopodium quinoa* Willd.) cultivars. *Iran. J. Field Crop Sci.* **2017**, *48*, 89–100. [CrossRef]
204. Strenske, A.; Vasconcelos, E.S.D.; Egewarth, V.A.; Herzog, N.F.M.; Malavasi, M.D.M. Responses of quinoa (*Chenopodium quinoa* willd.) seeds stored under different germination temperatures. *Acta Sci. Agron.* **2017**, *39*, 83–88. [CrossRef]
205. Bois, J.F.; Winkel, T.; Lhomme, J.P.; Raffaillac, J.P.; Rocheteau, A. Response of some Andean cultivars of quinoa (*Chenopodium quinoa* Willd.) to temperature: Effects on germination, phenology, growth and freezing. *Eur. J. Agron.* **2006**, *25*, 299–308. [CrossRef]
206. Bertero, H.D. Response of developmental processes to temperature and photoperiod in quinoa (*Chenopodium quinoa* Willd.). *Food Rev. Int.* **2003**, *19*, 87–97. [CrossRef]
207. Bertero, D.; Medan, D.; Hall, A.J. Changes in apical morphology during floral initiation and reproductive development in quinoa (*Chenopodium quinoa* Willd.). *Ann. Bot.* **1996**, *78*, 317–324. [CrossRef]
208. Bonifacio, A. Interspecific and Intergeneric Hybridization in Chenopod Species. Master's Thesis, Brigham Young University, Provo, UT, USA, 1995.
209. Lesjak, J.; Calderini, D.F. Increased night temperature negatively affects grain yield, biomass and grain number in Chilean quinoa. *Front. Plant Sci.* **2017**, *8*, 352. [CrossRef] [PubMed]
210. Bunce, J.A. Variation in yield responses to elevated CO2 and a brief high temperature treatment in quinoa. *Plants* **2017**, *6*, 26. [CrossRef] [PubMed]
211. Hinojosa, L.; Matanguihan, J.; Murphy, K. Effect of high temperature on pollen morphology, plant growth and seed yield in quinoa (*Chenopodium quinoa* Willd.). *J. Agron. Crop Sci.* **2018**. [CrossRef]
212. Bunce, J.A. Thermal acclimation of the temperature dependence of the VCmax of Rubisco in quinoa. *Photosynthetica* **2018**, 1171–1176. [CrossRef]
213. Sanabria, K.; Lazo, H. Aclimatación a la alta temperatura y tolerancia al calor (TL50) en 6 variedades de *Chenopodium quinoa*. *Rev. Peru. Biol.* **2018**, *25*, 147–152. [CrossRef]
214. Tashi, G.; Zhan, H.; Xing, G.; Chang, X.; Zhang, H.; Nie, X.; Ji, W.; Tashi, G.; Zhan, H.; Xing, G.; et al. Genome-wide identification and expression analysis of heat shock transcription factor family in *Chenopodium quinoa* Willd. *Agronomy* **2018**, *8*, 103. [CrossRef]
215. Bertero, H.D.; King, R.W.; Hall, A.J. Photoperiod-sensitive development phases in quinoa (*Chenopodium quinoa* Willd.). *Field Crop. Res.* **1999**, *60*, 231–243. [CrossRef]
216. Bertero, H.D.; King, R.W.; Hall, A.J. Modelling photoperiod and temperature responses of flowering in quinoa (*Chenopodium quinoa* Willd.). *Field Crop. Res.* **1999**, *63*, 19–34. [CrossRef]
217. Bertero, H.D.; King, R.W.; Hall, A.J. Photoperiod and temperature effects on the rate of leaf appearance in quinoa (*Chenopodium quinoa*). *Funct. Plant Biol.* **2000**, *27*, 349–356. [CrossRef]
218. Bertero, H. Effects of photoperiod, temperature and radiation on the rate of leaf appearance in quinoa (*Chenopodium quinoa* Willd.) under field conditions. *Ann. Bot.* **2001**, *87*, 495–502. [CrossRef]
219. Delatorre-Herrera, J.; Gonzalez, J.L.; Martinez, E. Efecto del fotoperiodo y la temperartura sobre la concentracion de saponina en tres variedades de quinua (*Chenopodium quinoa*) provenientes de tres latitudes. In *V Congreso Mundial de la Quinua y II Simposio Internacional de Granos Andinos*; V Congreso Mundial de la Quinua y II Simposio Internacional de Granos Andinos: San Salvador de Jujuy, Argentina, 2015; p. 70.
220. Curti, R.N.; Sajama, J.; Ortega-Baes, P. Setting conservation priorities for Argentina's pseudocereal crop wild relatives. *Biol. Conserv.* **2017**, *209*, 349–355. [CrossRef]
221. Jellen, E.N.; Kolano, B.A.; Sederberg, M.C.; Bonifacio, A.; Maughan, P.J. Chenopodium. In *Wild Crop Relatives: Genomic and Breeding Resources*; Kole, C., Ed.; Springer: Berlin/Heidelberg, Germany, 2011; pp. 35–61, ISBN 978-3-642-14386-1.

222. Wilson, H.D. Allozyme variation and morphological relationships of *Chenopodium hircinum* (s.l.). *Syst. Bot.* **1988**, *13*, 215. [CrossRef]
223. Wilson, H.; Manhart, J. Crop/weed gene flow: *Chenopodium quinoa* Willd. and C. berlandieri Moq. *Theor. Appl. Genet.* **1993**, *86*, 642–648. [CrossRef] [PubMed]
224. Müller-Xing, R.; Xing, Q.; Goodrich, J. Footprints of the sun: Memory of UV and light stress in plants. *Front. Plant Sci.* **2014**, *5*. [CrossRef]
225. Yao, X.; Liu, Q. Responses in growth, physiology and nitrogen nutrition of dragon spruce (*Picea asperata*) seedlings of different ages to enhanced ultraviolet-B. *Acta Physiol. Plant.* **2007**, *29*, 217–224. [CrossRef]
226. Lindroth, R.L.; Hofman, R.W.; Campbell, B.D.; McNabb, W.C.; Hunt, D.Y. Population differences in *Trifolium repens* L. response to ultraviolet-B radiation: Foliar chemistry and consequences for two lepidopteran herbivores. *Oecologia* **2000**, *122*, 20–28. [CrossRef] [PubMed]
227. Horak, E.; Farré, E.M. The regulation of UV-B responses by the circadian clock. *Plant Signal. Behav.* **2015**, *10*. [CrossRef] [PubMed]
228. Jenkins, G.I. The UV-B photoreceptor UVR8: From structure to physiology. *Plant Cell* **2014**, *26*, 21–37. [CrossRef] [PubMed]
229. Rizzini, L.; Favory, J.-J.; Cloix, C.; Faggionato, D.; O'Hara, A.; Kaiserli, E.; Baumeister, R.; Schäfer, E.; Nagy, F.; Jenkins, G.I.; et al. Perception of UV-B by the *Arabidopsis* UVR8 protein. *Science* **2011**, *332*, 103–106. [CrossRef] [PubMed]
230. Robson, T.M.; Klem, K.; Urban, O.; Jansen, M.A.K. Re-interpreting plant morphological responses to UV-B radiation. *Plant Cell Environ.* **2015**, *38*, 856–866. [CrossRef] [PubMed]
231. McKenzie, R.L.; Liley, J.B.; Björn, L.O. UV radiation: Balancing risks and benefits. *Photochem. Photobiol.* **2009**, *85*, 88–98. [CrossRef] [PubMed]
232. Palenque, E.; Andrade, M.; González, J.A.; Hilal, M.; Prado, F.E. Efectos de la radiación ultravioleta sobre la quinoa (*Chenopodium quinoa* Willd.). *Rev. Boliv. Física* **1997**, *3*, 120–128.
233. Sircelj, M.R.; Rosa, M.; Parrado, M.F.; González, J.A.; Hilal, M.; Prado, F.E. Ultrastructural and metabolic changes induced by UV-B radiation in cotyledons of quinoa. *Biocell* **2002**, *26*, 180.
234. Hilal, M.; Parrado, M.F.; Rosa, M.; Gallardo, M.; Orce, L.; Massa, E.M.; González, J.A.; Prado, F.E. Epidermal lignin deposition in quinoa cotyledons in response to UV-B radiation. *Photochem. Photobiol.* **2004**, *79*, 205–210. [CrossRef]
235. González, J.A.; Rosa, M.; Parrado, M.F.; Hilal, M.; Prado, F.E. Morphological and physiological responses of two varieties of a highland species (*Chenopodium quinoa* Willd.) growing under near-ambient and strongly reduced solar UV–B in a lowland location. *J. Photochem. Photobiol. B* **2009**, *96*, 144–151. [CrossRef] [PubMed]
236. Perez, M.L.; González, J.A.; Prado, F.E. Efectos de la radiación ultravioleta B (UVB) sobre diferentes variedades de Quinoa: I. Efectos sobre la morfología en condiciones controladas. *Bol. Soc. Argent. Bot.* **2015**, *50*, 337–347.
237. Prado, F.E.; González, J.A.; Perez, M.L. Efectos de la radiación ultravioleta B (UV-B) sobre diferentes variedades de Quinoa: II.- efectos sobre la síntesis de pigmentos fotosintéticos, protectores y azúcares solubles en condiciones controladas. *Bol. Soc. Argent. Bot.* **2016**, *51*, 665–673. [CrossRef]
238. Reyes, T.H.; Scartazza, A.; Castagna, A.; Cosio, E.G.; Ranieri, A.; Guglielminetti, L. Physiological effects of short acute UVB treatments in *Chenopodium quinoa* Willd. *Sci. Rep.* **2018**, *8*, 371. [CrossRef] [PubMed]
239. Bhargava, A.; Shukla, S.; Srivastava, J.; Singh, N.; Ohri, D. Genetic diversity for mineral accumulation in the foliage of *Chenopodium* spp. *Sci. Hortic.* **2008**, *118*, 338–346. [CrossRef]
240. Jacobsen, S.-E.; Monteros, C.; Corcuera, L.J.; Bravo, L.A.; Christiansen, J.L.; Mujica, A. Frost resistance mechanisms in quinoa (*Chenopodium quinoa* Willd.). *Eur. J. Agron.* **2007**, *26*, 471–475. [CrossRef]
241. Jayme-Oliveira, A.; Júnior, R.; Quadros, W.; Ramos, M.L.G.; Ziviani, A.C.; Jakelaitis, A.; Jayme-Oliveira, A.; Júnior, R.; Quadros, W.; Ramos, M.L.G.; et al. Amaranth, quinoa, and millet growth and development under different water regimes in the Brazilian Cerrado. *Pesqui. Agropecu. Bras.* **2017**, *52*, 561–571. [CrossRef]
242. Rosa, M.; Hilal, M.; González, J.A.; Prado, F.E. Low-temperature effect on enzyme activities involved in sucrose–starch partitioning in salt-stressed and salt-acclimated cotyledons of quinoa (*Chenopodium quinoa* Willd.) seedlings. *Plant Physiol. Biochem.* **2009**, *47*, 300–307. [CrossRef] [PubMed]
243. Ceccato, D.; Bertero, D.; Batlla, D.; Galati, B. Structural aspects of dormancy in quinoa (*Chenopodium quinoa*): Importance and possible action mechanisms of the seed coat. *Seed Sci. Res.* **2015**, *25*, 267–275. [CrossRef]

244. Ceccato, D.V.; Bertero, H.D.; Batlla, D. Environmental control of dormancy in quinoa (*Chenopodium quinoa*) seeds: Two potential genetic resources for pre-harvest sprouting tolerance. *Seed Sci. Res.* **2011**, *21*, 133–141. [CrossRef]
245. Thomas, E.C.; Lavkulich, L.M. Community considerations for quinoa production in the urban environment. *Can. J. Plant Sci.* **2014**, *95*, 397–404. [CrossRef]
246. Scoccianti, V.; Bucchini, A.E.; Iacobucci, M.; Ruiz, K.B.; Biondi, S. Oxidative stress and antioxidant responses to increasing concentrations of trivalent chromium in the Andean crop species *Chenopodium quinoa* Willd. *Ecotoxicol. Environ. Saf.* **2016**, *133*, 25–35. [CrossRef] [PubMed]

© 2018 by the authors. Licensee MDPI, Basel, Switzerland. This article is an open access article distributed under the terms and conditions of the Creative Commons Attribution (CC BY) license (http://creativecommons.org/licenses/by/4.0/).

Review

Towards an Understanding of the Molecular Basis of Nickel Hyperaccumulation in Plants

Llewelyn van der Pas and Robert A. Ingle *

Department of Molecular and Cell Biology, University of Cape Town, Rondebosch 7701, South Africa; vpslle001@myuct.ac.za
* Correspondence: robert.ingle@uct.ac.za; Tel.: +27-21-650-2408

Received: 6 December 2018; Accepted: 31 December 2018; Published: 4 January 2019

Abstract: Metal hyperaccumulation is a rare and fascinating phenomenon, whereby plants actively accumulate high concentrations of metal ions in their above-ground tissues. Enhanced uptake and root-to-shoot translocation of specific metal ions coupled with an increased capacity for detoxification and sequestration of these ions are thought to constitute the physiological basis of the hyperaccumulation phenotype. Nickel hyperaccumulators were the first to be discovered and are the most numerous, accounting for some seventy-five percent of all known hyperaccumulators. However, our understanding of the molecular basis of the physiological processes underpinning Ni hyperaccumulation has lagged behind that of Zn and Cd hyperaccumulation, in large part due to a lack of genomic resources for Ni hyperaccumulators. The advent of RNA-Seq technology, which allows both transcriptome assembly and profiling of global gene expression without the need for a reference genome, has offered a new route for the analysis of Ni hyperaccumulators, and several such studies have recently been reported. Here we review the current state of our understanding of the molecular basis of Ni hyperaccumulation in plants, with an emphasis on insights gained from recent RNA-Seq experiments, highlight commonalities and differences between Ni hyperaccumulators, and suggest potential future avenues of research in this field.

Keywords: nickel; hyperaccumulation; serpentine; RNA-Seq; IREG; ferroportin; ZIP; histidine

1. Introduction

Nickel is the most recent element to be classified as essential for plant growth [1], and to date, its only known biochemical function in plants is in the active site of the enzyme urease, which contains a bi-nickel center [2]. Presumably reflecting this, the requirement of plants for Ni is extremely low, and Ni deficiency, correspondingly rare. Higher plants typically contain Ni concentrations in the range of 0.5–10 mg kg^{-1} DW [3], and concentrations in excess of 10–50 mg kg^{-1} DW (depending on the plant species concerned) are associated with Ni toxicity effects [4]. These can include inhibition of photosynthesis, nitrogen assimilation, mitosis, and enzyme activity as well as DNA damage and the generation of reactive oxygen species [3,4]. While the majority of plants attempt to exclude excess Ni from their photosynthetically active tissues [5], a small group of plants known as Ni hyperaccumulators, actively accumulate Ni to concentrations in excess of 1000 mg kg^{-1} DW in their shoot tissues with no apparent toxicity effects [6].

Several hypotheses have been put forward to explain the adaptive function of metal hyperaccumulation in plants, of which the "elemental defence" hypothesis is most favored. This hypothesis, formulated by Boyd and Martens [7], proposes that the elevated concentrations of sequestered metal ions protect hyperaccumulators against attack by herbivores and pathogens. Feeding choice experiments carried out on several Ni hyperaccumulators have provided support for this hypothesis by demonstrating that herbivores opt to feed on leaf material with low Ni contents and that their fitness is reduced when forced to consume material with high Ni contents [8–10]. Ni hyperaccumulation has also been shown to reduce susceptibility to both bacterial and fungal pathogens in *Streptanthus polygaloides* [11], though it was associated with increased susceptibility of this species to Turnip mosaic virus [12]. Interestingly, the elemental defense provided by Ni has recently been shown to extend to the seeds produced by hyperaccumulators. Mortality rates of the generalist seed herbivore *Tribolium confusum* were significantly higher when fed seeds of *S. polygaloides* (containing 300 µg Ni g^{-1} DW) versus *Streptanthus insignis* (5 µg Ni g^{-1} DW), with an artificial diet experiment confirming that Ni concentrations > 240 µg Ni g^{-1} DW are toxic to this species [13].

There are currently 721 plant species in the global hyperaccumulator database, of which 532 are listed as Ni hyperaccumulators [14]. The taxonomic distribution of these species indicates that Ni hyperaccumulation has evolved multiple times. That said, 340 Ni hyperaccumulator species are found in just five families; Phyllanthaceae (118 species), Brassicaceae (87 species, 61 from the genus *Alyssum* in SE Europe and the Middle East), Cunoniaceae (48 species, all from New Caledonia), Asteraceae (45 species) and Euphorbiaceae (42 species, predominantly from Cuba). Since the publication of this database, several new Ni hyperaccumulators have been identified including *Senecio conrathii* from South Africa [15] and *Phyllanthus rufuschaneyi* from Borneo [16]. The preponderance of Ni hyperaccumulators among known metal hyperaccumulating plants likely reflects the fact that the serpentine soils with which they are associated are the most widespread metalliferous soils on a global scale [17]. Serpentine soils are derived from ultramafic rocks, and typically display a high Mg to Ca ratio, low levels of macronutrients and high levels of Ni, Co, and Cr [18,19]. The ultramafic regions of Turkey, Cuba, and New Caledonia, in particular, are regarded as "hotspots" of Ni hyperaccumulator biodiversity (see Reference [17] for a recent review of the ecology and distribution of hyperaccumulators).

The majority of Ni hyperaccumulators are restricted to serpentine soils (so-called obligate hyperaccumulators) where they can form almost pure species stands [17]. In contrast, the ranges of facultative hyperaccumulators extend outside of ultramafic outcrops [20]. While most facultative hyperaccumulators accumulate Ni whenever found on serpentine soils, some exceptions have been reported. For example, *Senecio coronatus* (Asteraceae) and *Alyssum sibiricum* (Brassicaceae) accumulate Ni at some but not all serpentine sites within their ranges [21,22]. Substantial intraspecific variation in Ni content can result from environmental factors, as is the case for *Pimelea leptospermoides* where shoot Ni contents ranging from 13 to 2873 mg kg^{-1} DW have been attributed to variation in total soil Ni content and pH [23]. However, in *S. coronatus*, this phenotypic variation has a genetic basis; plants from hyperaccumulator and non-accumulator populations have different root ultra-structures, and the accumulation phenotype of a given population persists when the plants are grown on a common soil substrate [24,25].

Like other metal hyperaccumulators, Ni hyperaccumulating plants display enhanced uptake, root to shoot translocation, and ability to detoxify and sequester Ni than non-accumulator species [26] (Figure 1). However, the molecular basis of these processes, and notably the transporters involved, is not well understood. In part, this is because the majority of research efforts aiming to uncover the molecular basis of metal hyperaccumulation have been directed at two Zn/Cd accumulators, *Arabidopsis halleri* and *Noccaea caerulescens* [20]. This is a consequence of their relatively recent divergence from *Arabidopsis thaliana*, which allowed the use of the genomic tools developed for this model plant. Expression profiling experiments have shown that these two hyperaccumulators have constitutively elevated expression of genes involved in the uptake, chelation, and xylem loading of Zn/Cd in comparison to related non-accumulators [27–29]. While it has been assumed that the same must hold true in Ni hyperaccumulators, the lack of genomic resources for these species has made this difficult to determine.

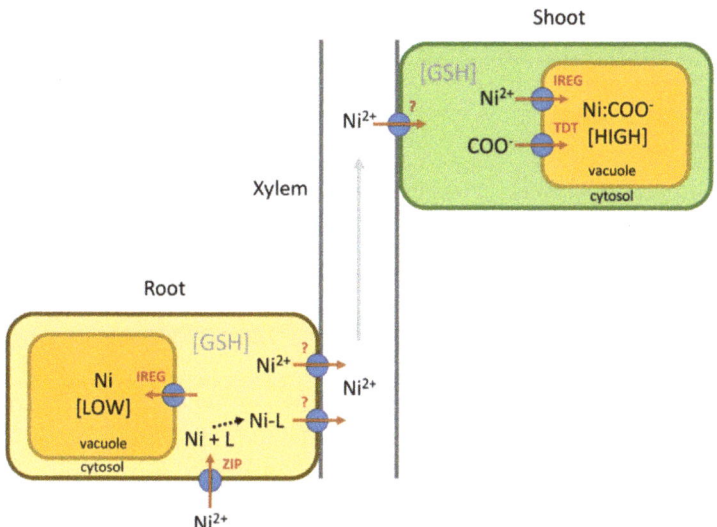

Figure 1. A general model for nickel hyperaccumulation in plants. Enhanced uptake of the Ni^{2+} cation from soil may be driven by high constitutive expression of poorly selective ZRT-IRT-like (ZIP) transporters responsible for Fe or Zn uptake. Chelation of Ni by an appropriate ligand (L) in the root cytosol is thought to prevent cytotoxicity. The identity and universality of the ligands used are an ongoing topic of debate, but histidine or nicotianamine perform this role in some Brassicaceae Ni hyperaccumulators. Formation of a Ni-ligand (Ni-L) complex may also impede the vacuolar sequestration of Ni in root tissues by tonoplast localized iron-regulated/ferroportin (IREG/FPN) transporters. Whether Ni is loaded into the xylem as the free cation or as a Ni-ligand complex is unclear, and the transporter(s) involved have not been identified, but the majority of Ni is present as the free cation in xylem sap. The transporter(s) involved in xylem unloading into shoot cells is also unknown. Ni is accumulated primarily in the shoot epidermis in most species, with the vacuole the major subcellular site of Ni sequestration. Constitutively high expression of IREG/FPN transporters has been reported in Ni hyperaccumulators versus related non-accumulators across four families, and two of these transporters have been shown to drive vacuolar sequestration of Ni. In the vacuole Ni is complexed by carboxylic acids (COO^-), with Ni-citrate or Ni-malate the predominant complexes identified to date. Ni hyperaccumulating *Senecio coronatus* plants display greatly increased expression of the tonoplast dicarboxylate transporter (TDT) compared to non-accumulators. Ni hyperaccumulators have enhanced capacity for the detoxification of reactive oxygen species, which may involve elevated concentrations of glutathione (GSH), flavonoids or increased activities of anti-oxidant enzymes.

This situation has changed recently with the advent of RNA-Seq technology which allows the de novo assembly of transcriptomes and quantification of transcript abundance in the absence of any prior sequence information, and to date, four such studies have been published on Ni hyperaccumulators. Merlot et al. [30] carried out de novo transcriptome assembly of the New Caledonian hyperaccumulator *Psychotria gabriellae*, while Halimaa et al. [31] performed a comparative analysis of gene expression in the roots of three accessions of *N. caerulescens*, including one from a Ni hyperaccumulating serpentine population. While representing significant milestones in the study of Ni hyperaccumulation, both studies had limitations. The first was restricted to a single species meaning that the gene expression values derived from it cannot readily be used to identify genes expressed at higher levels in Ni hyperaccumulators versus non-accumulators. Interpretation of the second study is complicated by the fact that the three *N. caerulescens* populations also vary in their capacity to accumulate Zn/Cd, are geographically distant and grow on very different soils. Subsequently, two comparative RNA-Seq studies have been reported. Meier et al. [24] performed a comparative RNA-Seq analysis of Ni hyperaccumulating and non-accumulating serpentine populations of *S. coronatus* to identify candidate genes that may underpin the Ni hyperaccumulation phenotype. A large-scale RNA-Seq study comparing seven pairs of related Ni hyperaccumulating and non-accumulating species across five families (Brassicaceae, Rubiaceae, Cunoniaceae, Salicaceae and Euphorbiaceae) from Cuba, New Caledonia and France has recently been reported [32]. In addition to facilitating the identification of changes in gene expression that are common across multiple independent evolutionary origins of Ni hyperaccumulation, this study reported the first gene knockdown experiment in a Ni hyperaccumulator species. Here we review what is currently known about the molecular basis of the physiological processes underlying Ni hyperaccumulation in plants, with an emphasis on newly published results, and highlight the main challenges ahead and questions still to be answered.

2. Ni Uptake

To date, there is no evidence to suggest the existence of a Ni-specific transporter in plants for the uptake of this cation from soil. Instead, Ni uptake by roots appears to be catalyzed by poorly selective cation transporters, notably members of the ZRT/IRT-like (ZIP) family [33,34]. In *A. thaliana*, IRT1 is expressed in the root epidermis in response to Fe deficiency and is the primary route of Fe uptake in strategy I plants. However, IRT1 has a broad substrate specificity towards divalent cations and also transports Zn, Co, Cd, Mn, and Ni [35,36]. IRT1 is thought to be the primary route for Ni uptake in *A. thaliana*, at least under Fe limiting conditions. In wild-type plants, root Ni concentrations increased from one µmol g^{-1} FW under Fe-replete conditions to three µmol g^{-1} FW under Fe-deficiency, but remained unchanged in *irt1* mutants [32]. A more modest increase (less than 40%) in root Ni concentrations has also been reported under Zn-deficient conditions in *A. thaliana*, though the transporter(s) involved in this uptake process are unknown. The root Zn transporter ZIP3 is not involved as a *zip3* mutant actually accumulated higher levels of Ni under Zn-deficiency than did wild-type plants [33].

In line with the hypothesis that Ni uptake by roots occurs via non-selective transporters, inhibition of Ni accumulation by Zn and Fe has been reported in Ni hyperaccumulators. For example, supplementation of hydroponic media with equimolar Ni and Zn concentrations led to a significant reduction in shoot Ni content in the Zn/Ni hyperaccumulators *Noccaea pindicum* and *Noccaea alpinum* var. *sylvium* compared to plants treated with Ni only, while no inhibitory effect of Ni on Zn accumulation was observed [37]. Similarly, the presence of equimolar Ni and Zn concentrations had no effect on Zn uptake but resulted in an 80 to 90% reduction in Ni uptake in a serpentine population of *N. caerulescens* [38]. Such data suggest that Ni uptake in the *Noccaea* Ni hyperaccumulators is mediated via one or more Zn transporters.

In contrast, a recent study of Ni uptake kinetics in two *Alyssum* Ni hyperaccumulators, *Alyssum inflatum* and *Alyssum bracteatum*, found no evidence for inhibition of Ni uptake by equimolar concentrations of Zn [39]. Instead, in *A. bracteatum*, Ni uptake was modulated by the concentration of Fe present in hydroponic media. Plants grown in the absence of added Fe for one week displayed a higher V_{max} for Ni uptake, while those exposed to excess Fe had a higher K_m for Ni uptake [39]. These results are consistent with Ni uptake via an Fe-deficiency induced Fe transporter, such as IRT1, in this species. In stark contrast, the kinetics of Ni uptake in *A. inflatum* were unaffected by Fe concentration in the nutrient solution (while those of Fe were not) suggesting that Ni uptake does not occur via a Fe transporter in this species [39]. A subsequent study employing the Ca channel inhibitor verapamil lead to a significant decrease in Ni uptake in both species [40] which the authors suggest indicates that these may be a major route of Ni uptake. However, it is not possible to rule out pleiotrophic effects of verapamil on Ni uptake, and further experimental work is required to test this hypothesis.

Increased expression of several members of the ZIP family has been reported in two comparative RNA-seq studies of Ni hyperaccumulators. *S. coronatus* Ni hyperaccumulators display high expression of a *ZIP* transporter, putatively annotated as *IRT1* based on phylogenetic analysis, in their root tissues while non-accumulating serpentine plants do not [24]. Increased expression of transcripts putatively annotated as *IRT1* and *ZIP10* has also been reported in roots of serpentine versus non-serpentine populations of *N. caerulescens* [31]. However, annotation of ZIP transporters based solely on sequence similarity to *A. thaliana* ZIPs is problematic due to the extensive gene duplication that has occurred in the ZIP family. For example, in *A. thaliana* IRT1 is one of five paralogues which share high sequence similarity but have different functions.

3. Chelation and Xylem Loading of Ni

Chelation of Ni ions in the cytosol by ligands is thought to be an important tolerance mechanism by which hyperaccumulators prevent metal toxicity and growth impairment [41]. However, the nature of the ligands involved, and their relative importance remains a matter of some debate. The amino acid histidine (His) has been implicated in the tolerance and root-to-shoot transport of Ni in several hyperaccumulators from the Brassicaceae. *Noccaea goesingense* and *Alyssum lesbiacum* display constitutively elevated levels of free His in their root tissues compared to related non-accumulators [42,43], which in *A. lesbiacum*, at least, is due to elevated expression of the first enzyme in the His biosynthetic pathway ATP-phosphoribosyl transferase (ATP-PRT) [42]. *A. thaliana* plants over-expressing ATP-PRT display both elevated His contents and tolerance to Ni [42,44,45].

Nicotianamine (NA), which is synthesized from three molecules of S-adenosylmethionine, has also been implicated as a Ni chelator in the genus *Noccaea*. A Ni-NA complex has been identified in *N. caerulescens* [46], and a strong positive correlation between shoot Ni and NA contents across seven species was observed by Callahan et al. [47]. Over-expression of NA synthase in *A. thaliana* and tobacco results in increased NA levels and Ni tolerance [48]. However, as is the case for His, there is little evidence to date to suggest that NA is a significant chelator of Ni outside the Brassicaceae.

The identity of the transport protein(s) responsible for the xylem loading of Ni remains elusive. Exogenous Ni elicits a dose-dependent increase in xylem His levels in several *Alyssum* hyperaccumulators [49,50] suggesting that His might also be involved in xylem loading of Ni. However, this does not occur in the non-accumulator *Alyssum montanum* [50] nor in transgenic *A. thaliana* plants with elevated free His levels due to over-expression of ATP-PRT [42]. This has been interpreted as indicating that Ni hyperaccumulators may possess a transport protein, lacking in other plant species, that loads the Ni-His complex into the xylem, though no such transporter has been identified to date. It is also clear that the majority of Ni in the xylem exists as the free cation in the *Alyssum* Ni hyperaccumulators [50,51], and that xylem Ni concentrations are an order of magnitude higher than those of His [51]. As His strongly promotes Ni uptake by root-derived tonoplast vesicles of the non-accumulator *Noccaea arvense* but has a modest repressive effect on vesicles from *N. caerulescens* [52], it has been suggested that His could promote root-to-shoot translocation of Ni by reducing vacuolar sequestration of Ni in root tissues of hyperaccumulators [52], though the mechanism by which this might occur is unknown. The observation that root to shoot translocation of Ni was enhanced in both *A. inflatum* and *A. bracteatum* under conditions of Fe deficiency suggests that xylem loading of Ni may be mediated by the same transporter(s) responsible for the loading of Fe [37], which is thought to be as the free Fe^{3+} cation. The identity of this transporter is unknown, but IREG1 a member of the iron-regulated/ferroportin (IREG/FPN) family which localizes to the plasma membrane and displays vasculature-specific expression has been proposed as a candidate [53], and might conceivably also transport Ni.

4. Vacuolar Sequestration of Nickel

The shoot epidermis is the primary site of Ni deposition in the majority of Ni hyperaccumulators studied to date [54,55]. The transporters responsible for the unloading of Ni (and Fe) from xylem into shoot cells are unknown but may include members of the ZIP family [56]. Interestingly, the *ZIP* transporter identified from *S. coronatus* displays differential expression in roots and shoots of Ni hyperaccumulators versus non-accumulator plants and is most highly expressed in shoots [24]. At a subcellular level, the vacuole is the primary site of Ni storage in shoot cells [55,57], and vacuolar ATPase-dependent Ni/proton antiport activity has been demonstrated in vacuoles derived from shoot cells of *A. lesbiacum* [58].

In *A. thaliana*, IREG2, another member of the IREG/FPN family, mediates the vacuolar sequestration of Ni inadvertently taken up by IRT1 in roots, as evidenced by the reduced Ni tolerance of *ireg2* mutants under Fe-deficiency conditions [36]. There is increasing evidence to suggest that vacuolar uptake of Ni may also be mediated by IREG/FPN transporters in Ni hyperaccumulators. Merlot et al. [30] identified a vacuolar-targeted IREG (PgIREG1) from *P. gabriellae* and demonstrated using RT-qPCR that it is expressed at higher levels in this species than in the non-accumulator *Psychotria semperflorens*. Heterologous expression of *PgIREG1* complemented the hypersensitivity of root growth to Ni and reduced root Ni content phenotypes of the *A. thaliana ireg2* mutant, and a PgIREG1:GFP fusion protein localized to the vacuolar membrane [30]. While *PgIREG1* is a functional orthologue of *IREG2*, its pattern of expression differs markedly; in *A. thaliana IREG2* expression is restricted to the root and occurs only under Fe deficiency, while in *P. gabriellae* high constitutive expression occurs in both root and shoot tissues [30]. A similar pattern of expression was observed for the putative *IREG* homologue identified in *S. coronatus* Ni hyperaccumulators [24].

The recent cross-species RNA-Seq study performed by de la Torre et al. [32] has revealed that elevated expression of *IREG* transporters in shoot tissues is a common feature of Ni hyperaccumulators from the Brassicaceae (*N. caerulescens*), Euphorbiaceae (*Leuococroton havanensis*) and Rubiaceae from both New Caledonia (*P. gabriellae*) and Cuba (*Psychotria grandis* and *Psychotria costivenia*) compared to related non-accumulators. Notably, this study also reported the first gene silencing experiments to be carried out in a Ni hyperaccumulator. Roots of *N. caerulescens* were transformed with *Rhizobium rhizogenes*, and gene knockdown of *IREG* achieved through the production of artificial miRNA. The resulting decrease in root Ni contents, together with the tonoplast localization of the transporter, support a role for *IREG* in the vacuolar sequestration of Ni [32].

While carboxylic acids, such as citrate and malate, have low stability constants with Ni at cytosolic pH, they are the predominant ligands for Ni in the acidic environment of the vacuole [59]. The carboxylic acid utilized appears to vary between Ni hyperaccumulators. High levels of citrate and the presence of a Ni-citrate complex were reported from shoot tissues of several New Caledonian Ni hyperaccumulators including *Pycnandra acuminata* [60,61]. In contrast, malate appears to be the primary ligand for Ni in shoot tissues of the Brassicaceae Ni hyperaccumulators [62,63]. Both Ni-citrate and Ni-malate complexes have been identified in the South African Asteraceae hyperaccumulators *Berkheya coddii* and *S. coronatus* [64], and Ni hyperaccumulating *S. coronatus* plants display greatly elevated expression of the tonoplast dicarboxylate transporter in their shoot tissues in comparison to non-accumulators [24].

5. Protection against Oxidative Damage

As it is not redox active, Ni cannot generate reactive oxygen species (ROS) directly via electron transfer but can disrupt the balance between formation and destruction of ROS during metabolism. This is thought to occur via direct binding of Ni to proteins or by displacement of essential cations from binding sites thereby leading to oxidative stress [65]. Exposure to Ni has been shown to result in increased ROS levels in Ni hyperaccumulators. For example H_2O_2 levels increased 3.6-fold in hairy root cultures of *Alyssum bertolonii* in response to 25 ppm Ni [66], while dichlorofluorescein-based imaging revealed that ROS levels increased in roots of both *Alyssum murale* (hyperaccumulator) and *A. montanum* (non-accumulator) when transferred from hydroponic media lacking added Ni to media containing 80 µM Ni [67]. Such experiments indicate that Ni hyperaccumulators do not have the ability to prevent Ni-induced ROS formation from occurring, and instead, it seems likely that they possess enhanced ROS detoxification capacity.

Whether Ni hyperaccumulators possess constitutively elevated anti-oxidant enzyme activities to deal with Ni-induced ROS relative to non-accumulators is debatable. Some studies have suggested that this is the case, for example, super-oxide dismutase (SOD) and catalase activities were determined to be 2.4 and 500 times greater respectively in *A. bertolonii* versus tobacco hairy root cultures [66]. In contrast, other studies have found little or no difference. For example, in a study comparing *A. inflatum*, *A. bracteatum* (Ni hyperaccumulators) and *Alyssum saxatile* (non-accumulator), SOD activities were less than two-fold higher in roots of the hyperaccumulators and identical in shoot tissues [68]. There is also no evidence from RNA-Seq experiments to suggest that mRNA levels of these anti-oxidant enzymes are elevated in hyperaccumulators versus non-accumulators [24,32]. It is apparent that the activity of anti-oxidant enzymes can be overcome by high concentrations of Ni, as demonstrated by the decline in SOD and catalase activities observed in *A. markgrafii* when grown hydroponically in the presence of 0.5 mM Ni, correlating with reduced biomass production in this species [69].

In addition to enzymatic anti-oxidants, plants also contain non-enzymatic anti-oxidants that scavenge ROS including glutathione (GSH) and ascorbate. A strong positive correlation between shoot Ni and GSH contents has been demonstrated in the genus *Noccaea* [70]. This has been attributed to elevated activity of serine acetyltransferase (SAT) in Ni hyperaccumulating members of this genus, which is required for the production of cysteine (Cys), a component of GSH. *A. thaliana* plants over-expressing SAT from *N. goesingense* contain elevated levels of Cys and GSH and display increased Ni tolerance [70]. The regeneration of reduced GSH from oxidized glutathione disulphide (GSSG) is catalyzed by the enzyme glutathione reductase, and *N. goesingense* has both an increased glutathione reductase (GR) activity and GSH:GSSG ratio in comparison to *A. thaliana* [70]. GR activities and GSH levels have also been shown to increase in response to Ni in the hyperaccumulators *A. inflatum*, *A. bracteatum*, and *A. markgrafii*, but this does not occur in the non-accumulator *A. saxatile* [68,69]. Together these data suggest that the increased size of the GSH pool, and the ability to maintain the GSH:GSSG ratio through the activity of GR may be an important component of Ni tolerance in hyperaccumulators, at least, in the Brassicaceae.

In the cross-species RNA-Seq experiments performed by de la Torre et al. [32], ten of the 33 clusters of orthologous genes more highly expressed in at least three of the hyperaccumulator versus non-accumulator pairwise species comparisons function in phenylpropanoid and flavonoid biosynthesis. Flavonoids have been suggested to function as scavengers of ROS generated during environmental stress, and their synthesis increases in plants under such conditions [71]. Support for this hypothesis has been provided by transgenic *A. thaliana* plants over-expressing *MYB12* which contain elevated levels of flavonoids, particularly anthocyanins, and display increased tolerance to oxidative and drought stress [72]. Whether ROS scavenging by flavonoids plays a role in Ni hyperaccumulators is unknown but is a worthwhile avenue for future research. Finally, Ni can cause DNA damage through direct oxidation of guanine residues, inhibition of DNA repair systems or by promoting the formation of ROS [73]. The observation that exogenous Ni caused DNA damage leading to reduced genomic integrity in *A. thaliana* but not in serpentine *N. caerulescens*, led to the suggestion that the capacity to maintain genome integrity in the presence of high levels of Ni is an important component of the hyperaccumulation phenotype [74]. In line with this hypothesis, *S. coronatus* hyperaccumulators have elevated expression of several genes involved in telomere maintenance, DNA repair and maintenance of repressive epigenetic marks in comparison to non-accumulators [24].

6. Conclusions and Potential Directions of Future Research

Ni hyperaccumulation has evolved multiple times in plants, but whether it has done so via the same route on each occasion is unknown. For some aspects of the Ni hyperaccumulation phenotype, there is evidence of convergent evolution, most notably in the high constitutive expression of *IREG/FPN* orthologues implicated in vacuolar sequestration of Ni in the shoot tissues of Ni hyperaccumulators from the Asteraceae, Brassicaceae, Euphorbiaceae, and Rubiaceae [24,30,32]. However other components, such as the role of His or NA as significant Ni-binding ligands, appear to be restricted to specific plant lineages. There is also evidence suggesting that the route by which Ni is taken up from the soil varies between hyperaccumulators within the same family, with Zn transporters implicated in *Noccaea* hyperaccumulators and Fe transporters in *Alyssum* hyperaccumulators [37–39]. It is possible that this implies that different ZIP transporters can be recruited during the evolution of Ni hyperaccumulation. Given the known role of the ZIP transporter IRT1 in Ni uptake in *A. thaliana* [34], increased expression of members of the ZIP family in roots of *N. caerulescens* and *S. coronatus* suggests that they might be involved in driving enhanced Ni uptake from the soil. However, the extensive expansion of the ZIP gene family in *A. thaliana* means that it is problematic to infer potential transport activities simply on the basis of sequence homology to *A. thaliana* ZIP transporters. For example, a putative IRT1 orthologue from barley was recently shown to transport Mn but not Fe [75]. Functional characterization of candidate Ni transport proteins to determine their actual substrate range is essential and could be accomplished through heterologous expression in yeast or by complementation of *A. thaliana* or yeast transporter mutants.

Better still would be to demonstrate that a candidate gene identified in a Ni hyperaccumulator is indeed required for Ni hypertolerance or accumulation in that species. This is essential if our understanding of the molecular basis of Ni hyperaccumulation is to progress beyond drawing correlations between mRNA or metabolite levels and Ni tolerance or accumulation in a hyperaccumulator species. To this end, the strategy of *R. rhizogenes* mediated transformation to effect gene silencing through artificial miRNA production recently used by de la Torre et al. [32] may be applicable to other Ni hyperaccumulators. Alternatively, CRISPR/Cas9-mediated mutation offers great potential to test gene function in Ni hyperaccumulators, and has recently been used in the Cd hyperaccumulator *Sedum plumbizincicola* to demonstrate that heavy metal ATPase 1 (HMA1) is required for Cd export from the chloroplast in order to prevent inhibition of photosynthesis [76].

Finally, what drives the elevated expression of the FPN/IREG and ZIP transporters implicated in Ni transport in Ni hyperaccumulators? Zn accumulation in both *N. caerulescens* and *A. halleri* results (at least in part) from elevated expression of the *HMA4* transporter responsible for xylem loading of Zn, leading to Zn deficiency in the root tissues of these species, and so to up-regulation of the transporters involved in Zn uptake [77,78]. In *A. thaliana*, IRT1 and IREG2 expression is regulated by the transcription factor FIT, which is itself activated in response to Fe deficiency [79]. *S. coronatus* hyperaccumulators display elevated expression of a putative *FIT* orthologue in comparison to non-accumulators, however this transcript was only detected in shoot tissues, while increased expression of the putative *IRT1* and *IREG* orthologues occurs in both roots and shoots [24]. Increased gene copy number and changes in promoter activity have been implicated in the increased *HMA4* expression observed in Zn hyperaccumulators [77,78]. Analysis of gene copy number and promoter DNA sequences of candidate Ni hyperaccumulation genes, coupled with yeast-one-hybrid screens to identify the transcription factors controlling their expression, will shed light on the regulation and evolution of this fascinating phenotype in plants.

Author Contributions: Writing—original draft preparation, L.v.d.P., R.A.I.; Writing—review and editing, R.A.I.

Funding: This research received no external funding.

Acknowledgments: Ongoing research on Ni hyperaccumulation in the lab of R.A.I. is funded by the University of Cape Town and National Research Foundation of South Africa.

Conflicts of Interest: The authors declare no conflict of interest.

References

1. Brown, P.H.; Welch, R.M.; Cary, E.E. Nickel: A micronutrient essential for higher plants. *Plant Physiol.* **1987**, *85*, 801–803. [CrossRef] [PubMed]
2. Balasubramanian, A.; Ponnuraj, K. Crystal structure of the first plant urease from jack bean: 83 years of journey from its first crystal to molecular structure. *J. Mol. Biol.* **2010**, *400*, 274–283. [CrossRef] [PubMed]
3. Chen, C.; Huang, D.; Liu, J. Functions and toxicity of nickel in plants: Recent advances and future prospects. *Clean Soil Air Water* **2009**, *37*, 304–313. [CrossRef]

4. Yusuf, M.; Fariduddin, Q.; Hayat, S.; Ahmad, A. Nickel: An overview of uptake, essentiality and toxicity in plants. *Bull. Environ. Contam. Toxicol.* **2011**, *86*, 41861. [CrossRef] [PubMed]
5. Hanikenne, M.; Nouet, C. Metal hyperaccumulation and hypertolerance: A model for plant evolutionary genomics. *Curr. Opin. Plant Biol.* **2011**, *14*, 252–259. [CrossRef] [PubMed]
6. Reeves, R. The hyperaccumulation of nickel by serpentine plants. In *The Vegetation of Ultramafic (Serpentine) Soils*; Baker, A.J.M., Proctor, J., Reeves, R.D., Eds.; Intercept Ltd.: Andover, UK, 1992; pp. 253–277.
7. Boyd, R.; Martens, S. The raison d'être for metal hyperaccumulation by plants. In *The Vegetation of Ultramafic (Serpentine) Soils*; Baker, A.J.M., Proctor, J., Reeves, R.D., Eds.; Intercept Ltd.: Andover, UK, 1992; pp. 279–289.
8. Boyd, R.S.; Davis, M.A.; Wall, M.A.; Balkwill, K. Nickel defends the South African hyperaccumulator *Senecio coronatus* (Asteraceae) against *Helix aspersa* (Mollusca: Pulmonidae). *Chemoecology* **2002**, *12*, 91–97. [CrossRef]
9. Boyd, R.S.; Martens, S.N. Nickel hyperaccumulated by *Thlaspi montanum* var. *montanum* is acutely toxic to an insect herbivore. *Oikos* **1994**, *70*, 21–25.
10. Boyd, R.S.; Moar, W.J. The defensive function of Ni in plants: Response of the polyphagous herbivore *Spodoptera exigua* (Lepidoptera: Noctuidae) to hyperaccumulator and accumulator species of *Streptanthus* (Brassicaceae). *Oecologia* **1999**, *118*, 218–224. [CrossRef]
11. Boyd, R.S.; Shaw, J.J.; Martens, S.N. Nickel hyperaccumulation defends *Streptanthus polygaloides* (Brassicaceae) against pathogens. *Am. J. Bot.* **1994**, *81*, 294–300. [CrossRef]
12. Davis, M.A.; Murphy, J.F.; Boyd, R.S. Nickel increases susceptibility of a nickel hyperaccumulator to Turnip mosaic virus. *J. Environ. Q.* **2001**, *30*, 85–90. [CrossRef]
13. Mincey, K.A.; Boyd, R.S. Elemental defense of nickel hyperaccumulator seeds against a generalist insect granivore. *Ecol. Res.* **2018**, *33*, 561–570. [CrossRef]
14. Reeves, R.D.; Baker, A.J.; Jaffré, T.; Erskine, P.D.; Echevarria, G.; van der Ent, A. A global database for plants that hyperaccumulate metal and metalloid trace elements. *New Phytol.* **2018**, *218*, 407–411. [CrossRef] [PubMed]
15. Siebert, S.J.; Schutte, N.C.; Bester, S.P.; Komape, D.M.; Rajakaruna, N. *Senecio conrathii* NE Br. (Asteraceae), a new hyperaccumulator of nickel from serpentinite outcrops of the Barberton Greenstone Belt, South Africa. *Ecol. Res.* **2018**, *33*, 651–658. [CrossRef]
16. Bouman, R.; Van Welzen, P.; Sumail, S.; Echevarria, G.; Erskine, P.D.; Van Der Ent, A. *Phyllanthus rufuschaneyi*: A new nickel hyperaccumulator from Sabah (Borneo Island) with potential for tropical agromining. *Bot. Stud.* **2018**, *59*, 9. [CrossRef] [PubMed]
17. Reeves, R.D.; van der Ent, A.; Baker, A.J. Global distribution and ecology of hyperaccumulator plants. In *Agromining: Farming for Metals*; Springer: Berlin, Germany, 2018; pp. 75–92.
18. Baker, A.J.M.; Brooks, R.R. Terrestrial higher plants which hyperaccumulate metallic elements. A review of their distribution, ecology and phytochemistry. *Biorecovery* **1989**, *1*, 81–126.
19. Proctor, J. Toxins, nutrient shortages and droughts: The serpentine challenge. *Trends Ecol. Evol.* **1999**, *14*, 334–335. [CrossRef]
20. Pollard, A.J.; Reeves, R.D.; Baker, A.J.M. Facultative hyperaccumulation of heavy metals and metalloids. *Plant Sci.* **2014**, *217–218*, 8–17. [CrossRef]
21. Boyd, R.S.; Davis, M.A.; Balkwill, K. Elemental patterns in Ni hyperaccumulating and non-hyperaccumulating ultramafic soil populations of *Senecio coronatus*. *S. Afr. J. Bot.* **2008**, *74*, 158–162. [CrossRef]
22. Reeves, R.D.; Adigüzel, N. Rare plants and nickel accumulators from Turkish serpentine soils, with special reference to *Centaurea* species. *Turk. J. Bot.* **2004**, *28*, 147–153.
23. Reeves, R.D.; Laidlaw, W.S.; Doronila, A.; Baker, A.J.M.; Batianoff, G.N. Erratic hyperaccumulation of nickel, with particular reference to the Queensland serpentine endemic *Pimelea leptospermoides*. *Aust. J. Bot.* **2015**, *63*, 119–127. [CrossRef]
24. Meier, S.K.; Adams, N.; Wolf, M.; Balkwill, K.; Muasya, A.M.; Gehring, C.A.; Bishop, J.M.; Ingle, R.A. Comparative RNA-seq analysis of nickel hyperaccumulating and non-accumulating populations of *Senecio coronatus* (Asteraceae). *Plant J.* **2018**, *95*, 1023–1038. [CrossRef] [PubMed]
25. Mesjasz-Przybyłowicz, J.; Barnabas, A.; Przybyłowicz, W. Comparison of cytology and distribution of nickel in roots of Ni-hyperaccumulating and non-hyperaccumulating genotypes of *Senecio coronatus*. *Plant Soil* **2007**, *293*, 61–78. [CrossRef]

26. Rascio, N.; Navari-Izzo, F. Heavy metal hyperaccumulating plants: How and why do they do it? And what makes them so interesting? *Plant Sci.* **2011**, *180*, 169–181. [CrossRef] [PubMed]
27. Becher, M.; Talke, I.N.; Krall, L.; Krämer, U. Cross-species microarray transcript profiling reveals high constitutive expression of metal homeostasis genes in shoots of the zinc hyperaccumulator *Arabidopsis halleri*. *Plant J.* **2004**, *37*, 251–268. [CrossRef]
28. Van de Mortel, J.E.; Villanueva, L.A.; Schat, H.; Kwekkeboom, J.; Coughlan, S.; Moerland, P.D.; van Themaat, E.V.L.; Koornneef, M.; Aarts, M.G. Large expression differences in genes for iron and zinc homeostasis, stress response, and lignin biosynthesis distinguish roots of *Arabidopsis thaliana* and the related metal hyperaccumulator *Thlaspi caerulescens*. *Plant Physiol.* **2006**, *142*, 1127–1147. [CrossRef]
29. Weber, M.; Harada, E.; Vess, C.; Roepenack-Lahaye, E.V.; Clemens, S. Comparative microarray analysis of *Arabidopsis thaliana* and *Arabidopsis halleri* roots identifies nicotianamine synthase, a ZIP transporter and other genes as potential metal hyperaccumulation factors. *Plant J.* **2004**, *37*, 269–281. [CrossRef]
30. Merlot, S.; Hannibal, L.; Martins, S.; Martinelli, L.; Amir, H.; Lebrun, M.; Thomine, S. The metal transporter PgIREG1 from the hyperaccumulator *Psychotria gabriellae* is a candidate gene for nickel tolerance and accumulation. *J. Exp. Bot.* **2014**, *65*, 1551–1546. [CrossRef] [PubMed]
31. Halimaa, P.; Lin, Y.-F.; Ahonen, V.H.; Blande, D.; Clemens, S.; Gyenesei, A.; Häikiö, E.; Kärenlampi, S.O.; Laiho, A.; Aarts, M.G. Gene expression differences between *Noccaea caerulescens* ecotypes help to identify candidate genes for metal phytoremediation. *Environ. Sci. Technol.* **2014**, *48*, 3344–3353. [CrossRef] [PubMed]
32. De la Torre, V.S.G.; Majorel-Loulergue, C.; Gonzalez, D.A.; Soubigou-Taconnat, L.; Rigaill, G.J.; Pillon, Y.; Barreau, L.; Thomine, S.; Fogliani, B.; Burtet-Sarramegna, V. Wide cross-species RNA-Seq comparison reveals a highly conserved role for Ferroportins in nickel hyperaccumulation in plants. *bioRxiv* **2018**, 420729. [CrossRef]
33. Nishida, S.; Kato, A.; Tsuzuki, C.; Yoshida, J.; Mizuno, T. Induction of nickel accumulation in response to zinc deficiency in *Arabidopsis thaliana*. *Int. J. Mol. Sci.* **2015**, *16*, 9420–9430. [CrossRef]
34. Nishida, S.; Tsuzuki, C.; Kato, A.; Aisu, A.; Yoshida, J.; Mizuno, T. AtIRT1, the primary iron uptake transporter in the root, mediates excess nickel accumulation in *Arabidopsis thaliana*. *Plant Cell Physiol.* **2011**, *52*, 1433–1442. [CrossRef] [PubMed]
35. Korshunova, Y.O.; Eide, D.; Clark, W.G.; Guerinot, M.L.; Pakrasi, H.B. The IRT1 protein from *Arabidopsis thaliana* is a metal transporter with a broad substrate range. *Plant Mol. Biol.* **1999**, *40*, 37–44. [CrossRef] [PubMed]
36. Schaaf, G.; Honsbein, A.; Meda, A.R.; Kirchner, S.; Wipf, D.; von Wirén, N. AtIREG2 encodes a tonoplast transport protein involved in iron-dependent nickel detoxification in *Arabidopsis thaliana* roots. *J. Biol. Chem.* **2006**, *281*, 25532–25540. [CrossRef] [PubMed]
37. Taylor, S.I.; Macnair, M.R. Within and between population variation for zinc and nickel accumulation in two species of *Thlaspi* (Brassicaceae). *New Phytol.* **2006**, *169*, 505–514. [CrossRef]
38. Assunção, A.; Martins, P.D.C.; De Folter, S.; Vooijs, R.; Schat, H.; Aarts, M. Elevated expression of metal transporter genes in three accessions of the metal hyperaccumulator *Thlaspi caerulescens*. *Plant Cell Environ.* **2001**, *24*, 217–226. [CrossRef]
39. Mohseni, R.; Ghaderian, S.M.; Ghasemi, R.; Schat, H. Differential effects of iron starvation and iron excess on nickel uptake kinetics in two Iranian nickel hyperaccumulators, *Odontarrhena bracteata* and *Odontarrhena inflata*. *Plant Soil* **2018**, *428*, 153–162. [CrossRef]
40. Mohseni, R.; Ghaderian, S.M.; Schat, H. Nickel uptake mechanisms in two Iranian nickel hyperaccumulators, *Odontarrhena bracteata* and *Odontarrhena inflata*. *Plant Soil* **2018**. [CrossRef]
41. Van Der Ent, A.; Callahan, D.L.; Noller, B.N.; Mesjasz-Przybylowicz, J.; Przybylowicz, W.J.; Barnabas, A.; Harris, H.H. Nickel biopathways in tropical nickel hyperaccumulating trees from Sabah (Malaysia). *Sci. Rep.* **2017**, *7*, 41861. [CrossRef]
42. Ingle, R.A.; Mugford, S.T.; Rees, J.D.; Campbell, M.M.; Smith, J.A.C. Constitutively high expression of the histidine biosynthetic pathway contributes to nickel tolerance in hyperaccumulator plants. *Plant Cell* **2005**, *17*, 2089–2106. [CrossRef]
43. Persans, M.W.; Yan, X.; Patnoe, J.-M.M.; Krämer, U.; Salt, D.E. Molecular dissection of the role of histidine in nickel hyperaccumulation in *Thlaspi goesingense* (Hálácsy). *Plant Physiol.* **1999**, *121*, 1117–1126. [CrossRef]

44. Rees, J.D.; Ingle, R.A.; Smith, J.A.C. Relative contributions of nine genes in the pathway of histidine biosynthesis to the control of free histidine concentrations in *Arabidopsis thaliana*. *Plant Biotechnol. J.* **2009**, *7*, 499–511. [CrossRef] [PubMed]
45. Wycisk, K.; Kim, E.J.; Schroeder, J.I.; Krämer, U. Enhancing the first enzymatic step in the histidine biosynthesis pathway increases the free histidine pool and nickel tolerance in *Arabidopsis thaliana*. *FEBS Lett.* **2004**, *578*, 128–134. [CrossRef] [PubMed]
46. Mari, S.; Gendre, D.; Pianelli, K.; Ouerdane, L.; Lobinski, R.; Briat, J.-F.; Lebrun, M.; Czernic, P. Root-to-shoot long-distance circulation of nicotianamine and nicotianamine-nickel chelates in the metal hyperaccumulator *Thlaspi caerulescens*. *J. Exp. Bot.* **2006**, *57*, 4111–4122. [CrossRef] [PubMed]
47. Callahan, D.L.; Kolev, S.D.; O'Hair, R.A.; Salt, D.E.; Baker, A.J. Relationships of nicotianamine and other amino acids with nickel, zinc and iron in *Thlaspi* hyperaccumulators. *New Phytol.* **2007**, *176*, 836–848. [CrossRef] [PubMed]
48. Kim, S.; Takahashi, M.; Higuchi, K.; Tsunoda, K.; Nakanishi, H.; Yoshimura, E.; Mori, S.; Nishizawa, N.K. Increased nicotianamine biosynthesis confers enhanced tolerance of high levels of metals, in particular nickel, to plants. *Plant Cell Physiol.* **2005**, *46*, 1809–1818. [CrossRef] [PubMed]
49. Kerkeb, L.; Krämer, U. The role of free histidine in xylem loading of nickel in *Alyssum lesbiacum* and *Brassica juncea*. *Plant Physiol.* **2003**, *131*, 716–724. [CrossRef] [PubMed]
50. Krämer, U.; Cotter-Howells, J.D.; Charnock, J.M.; Baker, A.J.; Smith, J.A.C. Free histidine as a metal chelator in plants that accumulate nickel. *Nature* **1996**, *379*, 635–638. [CrossRef]
51. Centofanti, T.; Sayers, Z.; Cabello-Conejo, M.I.; Kidd, P.; Nishizawa, N.K.; Kakei, Y.; Davis, A.P.; Sicher, R.C.; Chaney, R.L. Xylem exudate composition and root-to-shoot nickel translocation in *Alyssum* species. *Plant Soil* **2013**, *373*, 59–75. [CrossRef]
52. Richau, K.H.; Kozhevnikova, A.D.; Seregin, I.V.; Vooijs, R.; Koevoets, P.L.; Smith, J.A.C.; Ivanov, V.B.; Schat, H. Chelation by histidine inhibits the vacuolar sequestration of nickel in roots of the hyperaccumulator *Thlaspi caerulescens*. *New Phytol.* **2009**, *183*, 106–116. [CrossRef]
53. Morrissey, J.; Baxter, I.R.; Lee, J.; Li, L.; Lahner, B.; Grotz, N.; Kaplan, J.; Salt, D.E.; Guerinot, M.L. The ferroportin metal efflux proteins function in iron and cobalt homeostasis in Arabidopsis. *Plant Cell* **2009**, *21*, 3326–3338. [CrossRef]
54. Jaffré, T.; Reeves, R.D.; Baker, A.J.; Schat, H.; van der Ent, A. The discovery of nickel hyperaccumulation in the New Caledonian tree *Pycnandra acuminata* 40 years on: An introduction to a Virtual Issue. *New Phytol.* **2018**, *218*, 397–400. [CrossRef] [PubMed]
55. Küpper, H.; Lombi, E.; Zhao, F.J.; Wieshammer, G.; McGrath, S.P. Cellular compartmentation of nickel in the hyperaccumulators *Alyssum lesbiacum*, *Alyssum bertolonii* and *Thlaspi goesingense*. *J. Exp. Bot.* **2001**, *52*, 2291–2300. [CrossRef] [PubMed]
56. Visioli, G.; Gullì, M.; Marmiroli, N. *Noccaea caerulescens* populations adapted to grow in metalliferous and non-metalliferous soils: Ni tolerance, accumulation and expression analysis of genes involved in metal homeostasis. *Environ. Exp. Bot.* **2014**, *105*, 10–17. [CrossRef]
57. Krämer, U.; Pickering, I.J.; Prince, R.C.; Raskin, I.; Salt, D.E. Subcellular localization and speciation of nickel in hyperaccumulator and non-accumulator *Thlaspi* species. *Plant Physiol.* **2000**, *122*, 1343–1354. [CrossRef] [PubMed]
58. Ingle, R.; Fricker, M.; Smith, J. Evidence for nickel/proton antiport activity at the tonoplast of the hyperaccumulator plant *Alyssum lesbiacum*. *Plant Biol.* **2008**, *10*, 746–753. [CrossRef] [PubMed]
59. Bhatia, N.P.; Walsh, K.B.; Baker, A.J. Detection and quantification of ligands involved in nickel detoxification in a herbaceous Ni hyperaccumulator *Stackhousia tryonii* Bailey. *J. Exp. Bot.* **2005**, *56*, 1343–1349. [CrossRef]
60. Lee, J.; Reeves, R.D.; Brooks, R.R.; Jaffré, T. Isolation and identification of a citrato-complex of nickel from nickel-accumulating plants. *Phytochemistry* **1977**, *16*, 1503–1505. [CrossRef]
61. Lee, J.; Reeves, R.D.; Brooks, R.R.; Jaffré, T. The relation between nickel and citric acid in some nickel-accumulating plants. *Phytochemistry* **1978**, *17*, 1033–1035. [CrossRef]
62. Montargès-Pelletier, E.; Chardot, V.; Echevarria, G.; Michot, L.J.; Bauer, A.; Morel, J.-L. Identification of nickel chelators in three hyperaccumulating plants: An X-ray spectroscopic study. *Phytochemistry* **2008**, *69*, 1695–1709. [CrossRef]
63. Pelosi, P.; Fiorentini, R.; Galoppini, C. On the nature of nickel compounds in *Alyssum bertolonii* Desv.-II. *Agric. Biol. Chem.* **1976**, *40*, 1641–1642.

64. Montargès-Pelletier, E.; Mesjasz-Przybylowicz, J.; Barnabas, A.; Echevarria, G.; Briois, V.; Sechogela, T.P.; Przybylowicz, W.J. Do hyperaccumulators develop specific chelates for nickel transport and storage? The cases of *Senecio coronatus* and *Berkheya coddii*. In Proceedings of the 11th International Conference on Biogeochemistry of Trace Elements, Florence, Italy, 3–7 July 2011.
65. Sharma, S.S.; Dietz, K.-J. The relationship between metal toxicity and cellular redox imbalance. *Trends Plant Sci.* **2009**, *14*, 43–50. [CrossRef] [PubMed]
66. Boominathan, R.; Doran, P.M. Ni-induced oxidative stress in roots of the Ni hyperaccumulator, *Alyssum bertolonii*. *New Phytol.* **2002**, *156*, 205–215. [CrossRef]
67. Agrawal, B.; Czymmek, K.J.; Sparks, D.L.; Bais, H.P. Transient influx of nickel in root mitochondria modulates organic acid and reactive oxygen species production in nickel hyper-accumulator *Alyssum murale*. *J. Biol. Chem.* **2013**, *288*, 7351–7362. [CrossRef]
68. Ghaderian, S.M.; Ghasemi, R.; Heidari, H.; Vazirifar, S. Effects of Ni on superoxide dismutase and glutathione reductase activities and thiol groups: A comparative study between *Alyssum* hyperaccumulator and non-accumulator species. *Aust. J. Bot.* **2015**, *63*, 65–71. [CrossRef]
69. Stanisavljević, N.; Savić, J.; Jovanović, Ž.; Miljuš-Đukić, J.; Senćanski, J.; Simonović, M.; Radović, S.; Vinterhalter, D.; Vinterhalter, B. Fingerprinting of the antioxidant status in *Alyssum markgrafii* shoots during nickel hyperaccumulation in vitro. *Acta Physiol. Plant.* **2018**, *40*, 101. [CrossRef]
70. Freeman, J.L.; Persans, M.W.; Nieman, K.; Albrecht, C.; Peer, W.; Pickering, I.J.; Salt, D.E. Increased glutathione biosynthesis plays a role in nickel tolerance in *Thlaspi* nickel hyperaccumulators. *Plant Cell* **2004**, *16*, 2176–2191. [CrossRef] [PubMed]
71. Nakabayashi, R.; Saito, K. Integrated metabolomics for abiotic stress responses in plants. *Curr. Opin. Plant Biol.* **2015**, *24*, 10–16. [CrossRef]
72. Nakabayashi, R.; Yonekura-Sakakibara, K.; Urano, K.; Suzuki, M.; Yamada, Y.; Nishizawa, T.; Matsuda, F.; Kojima, M.; Sakakibara, H.; Shinozaki, K. Enhancement of oxidative and drought tolerance in Arabidopsis by overaccumulation of antioxidant flavonoids. *Plant J.* **2014**, *77*, 367–379. [CrossRef]
73. Cameron, K.S.; Buchner, V.; Tchounwou, P.B. Exploring the molecular mechanisms of nickel-induced genotoxicity and carcinogenicity: A literature review. *Rev. Environ. Health* **2011**, *26*, 81–92. [CrossRef]
74. Gullì, M.; Marchi, L.; Fragni, R.; Buschini, A.; Visioli, G. Epigenetic modifications preserve the hyperaccumulator *Noccaea caerulescens* from Ni geno-toxicity. *Environ. Mol. Mutagen.* **2018**, *59*, 464–475. [CrossRef]
75. Long, L.; Persson, D.P.; Duan, F.; Jørgensen, K.; Yuan, L.; Schjoerring, J.K.; Pedas, P.R. The iron-regulated transporter 1 plays an essential role in uptake, translocation and grain-loading of manganese, but not iron, in barley. *New Phytol.* **2018**, *217*, 1640–1653. [CrossRef]
76. Zhao, H.; Wang, L.; Zhao, F.J.; Wu, L.; Liu, A.; Xu, W. SpHMA1 is a chloroplast cadmium exporter protecting photochemical reactions in the Cd hyperaccumulator *Sedum plumbizincicola*. *Plant Cell Environ.* **2018**. [CrossRef] [PubMed]
77. Hanikenne, M.; Talke, I.N.; Haydon, M.J.; Lanz, C.; Nolte, A.; Motte, P.; Kroymann, J.; Weigel, D.; Krämer, U. Evolution of metal hyperaccumulation required cis-regulatory changes and triplication of *HMA4*. *Nature* **2008**, *453*, 391–395. [CrossRef]
78. Lochlainn, S.Ó.; Bowen, H.C.; Fray, R.G.; Hammond, J.P.; King, G.J.; White, P.J.; Graham, N.S.; Broadley, M.R. Tandem quadruplication of *HMA4* in the zinc (Zn) and cadmium (Cd) hyperaccumulator *Noccaea caerulescens*. *PLoS ONE* **2011**, *6*, e17814. [CrossRef] [PubMed]
79. Brumbarova, T.; Bauer, P.; Ivanov, R. Molecular mechanisms governing Arabidopsis iron uptake. *Trends Plant Sci.* **2015**, *20*, 124–133. [CrossRef]

© 2019 by the authors. Licensee MDPI, Basel, Switzerland. This article is an open access article distributed under the terms and conditions of the Creative Commons Attribution (CC BY) license (http://creativecommons.org/licenses/by/4.0/).

Article

Identification of Genomic Regions Contributing to Protein Accumulation in Wheat under Well-Watered and Water Deficit Growth Conditions

Ibrahim S. Elbasyoni [1,*], Sabah M. Morsy [1], Raghuprakash K. Ramamurthy [2] and Atef M. Nassar [3]

1. Crop Science Department, Faculty of Agriculture, Damanhour University, Damanhour 22516, Egypt; Sabahinunl@gmail.com
2. Department of Agronomy and Horticulture, University of Nebraska-Lincoln, Lincoln, NE 68583, USA; kasturiraghuprakash@gmail.com
3. Plant Protection Department, Faculty of Agriculture, Damanhour University, Damanhour 22516, Egypt; atef.nassar@dmu.edu.eg
* Correspondence: ibrahim.salah@agr.dmu.edu.eg; Tel.: +14028178156

Received: 6 June 2018; Accepted: 4 July 2018; Published: 11 July 2018

Abstract: Sustaining wheat production under low-input conditions through development and identifying genotypes with enhanced nutritional quality are two current concerns of wheat breeders. Wheat grain total protein content, to no small extent, determines the economic and nutritive value of wheat. Therefore, the objectives of this study are to identify accessions with high and low grain protein content (GPC) under well-watered and water-deficit growth conditions and to locate genomic regions that contribute to GPC accumulation. Spring wheat grains obtained from 2111 accessions that were grown under well-watered and water-deficit conditions were assessed for GPC using near-infrared spectroscopy (NIR). Results indicated significant influences of moisture, genotype, and genotype × environment interaction on the GPC accumulation. Furthermore, genotypes exhibited a wide range of variation for GPC, indicating the presence of high levels of genetic variability among the studied accessions. Around 366 (166 with high GPC and 200 with low GPC) wheat genotypes performed relatively the same across environments, which implies that GPC accumulation in these genotypes was less responsive to water deficit. Genome-wide association mapping results indicated that seven single nucleotide polymorphism (SNPs) were linked with GPC under well-watered growth conditions, while another six SNPs were linked with GPC under water-deficit conditions only. Moreover, 10 SNPs were linked with GPC under both well-watered and water-deficit conditions. These results emphasize the importance of using diverse, worldwide germplasm to dissect the genetic architecture of GPC in wheat and identify accessions that might be potential parents for high GPC in wheat breeding programs.

Keywords: wheat; grain protein content; water deficit; genome-wide association mapping

1. Introduction

Wheat (*Triticum aestivum* L.) is the food commodity for more than third of the world's population. Wheat grain is a rich source of starch (carbohydrate). Therefore wheat is primarily considered as a source of energy [1]. However, wheat grain contains also moderate amounts of dietary proteins which determines, to a large extent, both the end-use quality and wheat grain price [2]. Wheat grain total protein content (GPC) ranges from 9 to 15% of the dry weight [3,4]. Although, GPC depends primarily on the genotype; the environment and genotype × environment interaction also plays an essential role in grain protein accumulation [5].

Nitrogen fertilization is the critical environmental factor that affects protein accumulation; if nitrogen fertilization stays constant, increased yield often results in decreased protein content because of nitrogen dilution by the large biomass [6–9]. Furthermore, under stress conditions, grain protein content tends to be higher compared to either irrigated or nitrogen-limited conditions [10]. Water deficit increases grain total protein content, but it decreases grain yield [11]. Wheat genotypes with higher yield potential tend to have lower protein content and *vice versa* [12]. Several explanations for the negative relationship between grain protein content and yield have been proposed [13]. However, some wheat genotypes deviate from the previous relationship, i.e., they produce high yield and high grain protein content [14]. That deviation implies that the nitrogen supply to grains was increased, but it was not associated with a reduction in the grain yield [15].

Exploring genetic resources to identify wheat genotypes with high grain protein content is the most efficient way to improve the nutritional value of wheat grains [16]. Wheat breeders were successful in selecting genotypes with a high total protein content that were generated from cultivated materials such as "Atlas," "Atlas66" and "Nap Hall" [17]. Previous studies reported higher GPC in landraces and wild relatives compared to modern wheat genotypes [18,19]. A wild emmer (*Triticum turgidum var dicoccoides*) genotype was identified in Israel, i.e., "FA15-3" which was found to be able to accumulate 40% protein when given adequate nitrogen fertilization [20]. High grain protein content gene *GPC-B1* allele which was originally identified in wild emmer wheat, was transferred to a spring wheat genotype and increased grain total protein content by 3% [21]. The *GPC-B1* allele accelerates senescence and increases mobilization of nitrogen, zinc, and iron to the developing grains [22]. Thus, accessions containing this allele most likely will contain high protein as well as high iron and zinc [23]. However, most of the modern tetraploid and hexaploid wheat genotypes have lost a functional allele of *GPC-B1* [16]. During the last decade, several QTLs for GPC were mapped using association mapping (AM) and biparental populations on chromosomes 5A, 5D, 2D, 2B, 6A, 6B and 7A [24–31], that QTLs were validated and used in marker-assisted selection to improve GPC.

Marker-assisted selection (MAS) was defined as one of the promising avenues to improve wheat total protein content and grain yield [32]. The critical step in MAS is to identify molecular markers associated with desirable phenotypic traits using AM or biparental populations [33]. Association mapping (AM) can be applied to structured populations [34], thus incorporating a broad spectrum of germplasm is possible [34–36]. However, the successful application of association mapping requires comprehensive phenotypic and genotypic data. The dramatic decrease in the genotyping costs [37] in addition to the availability of high throughput phenotyping technologies such as Near-Infrared Spectroscopy (NIR) make AM a viable approach for large populations [38]. Furthermore, A robust sequence and annotation of the wheat genome are now available [39] with the latest developments in genomic technologies. This might allow researchers to identify new loci associated with GPC genes and dissect the genetic architecture of GPC in wheat.

Three strategies were adopted to select for high GPC and grain yield, i.e., selecting for high grain protein alone, selecting for high grain protein within highest yielding genotypes, and using an index to simultaneously select for both protein and yield [40]. In the current study, the most recent developments in genotyping and phenotyping technologies were applied to identify genomic regions associated with GPC and select accessions with high and low grain protein content using a worldwide collection of spring wheat accessions.

2. Materials and Methods

2.1. Plant Materials and Field Growth Conditions

Wheat grains obtained from 2111 wheat accessions (882 landraces; 493 breeding lines; 419 cultivars and 317 with uncertain category) were used in the current study. The accessions seeds were provided by the national small grains collection (NSGC) located in Aberdeen, ID, USA. The accessions were screened in Egypt during 2015/2016 and 2016/2017 growing seasons for total protein content under well-watered and water deficit conditions in Damanhour university experimental farm (30°45′19.4″ N, 30°29′4.8″ E). During the two growing seasons, drought stress was imposed by controlling irrigation during the reproductive stage in which plants were irrigated at 40% depletion of plant available water (PAW) (well-watered), or 80% PAW (water deficit). Well-watered and water deficit treatments were applied on two sublocations within the same experimental farm to facilitate the control of water application. For both sublocations, the wheat accessions were planted in two replicates using a randomized incomplete block design [41] in plots of four rows wide with 25 cm between rows and two meters long. The incomplete blocks consisted of 50 accessions in addition to three check cultivars, i.e., "Sids13", Gimmiza 9", and "Giza 168. The check cultivars were planted in each incomplete block.

2.2. Estimation of Grain Protein Content (GPC)

Grain protein content (% or g/100 g) was estimated using near-infrared spectroscopy (NIR) with a Perten DA7250 diode array NIR (Springfield, IL, USA). NIR is a nondestructive technique that complies with the ISO 12099 standard method. The measurements of GPC were done in the near infrared region 950–1650 nm and readings were processed in NetPlus software (Perten, Hägersten, Sweden), which includes validation calculation modules, such as calculations of bias, slope, and standard errors of prediction against the reference methods. However, for initial calibration of the Perten DA7250, the crude protein content of 100 wheat accessions was measured using the Kjeldahl method (Pelican Equipment's, Chennai, India). The correlation coefficient (r) between the calibration set and Perten DA7250 NIR readings was 0.964 for crude protein (% dry basis).

2.3. Single Nucleotide Polymorphism (SNP)

Wheat accessions included in this study were genotyped through the Triticeae Coordinated Agriculture Project (TCAP) using the Illumina iSelect 9 K (Illumina, Madison, WI, USA) wheat array [42] at the USDA-ARS genotyping laboratory in Fargo, ND, USA. The single nucleotide polymorphism (SNP) markers were filtered by removing SNPs with missing values >10% and minor allele frequency (MAF) <5%. The filtration step resulted in 5090 high-quality SNPs in which the missing values were imputed using random forest regression [43], which was applied using the MissForest R/package [44]. Then, the filtered and imputed SNP markers were used for the association mapping analysis, in which SNP markers were plotted in a Manhattan plot using "WNSP 2013 consensus map"; available on: (https://triticeaetoolbox.org/wheat/) according to Wang et al. [45].

2.4. Statistical Analysis

Analysis of variance was carried out by fitting the following model [46]:

$$Y_{ijlm} = \mu + E_i + EB_{(il)j} + G_m + EG_{im} + \varepsilon_{ijlm}$$

where Y_{ijlm} is the response measured on the $ijlm$ plot, μ is the overall mean, E_i is the effect of ith environment, $EB_{(il)j}$ is jth incomplete block nested within lth complete block and ith environment (random), G_m is the effect of mth accession, EG_{im} is the interaction effect among ith environment and mth accession, and ε_{ijlm} is the experimental error. Type III expected mean square estimation was conducted as follows:

Source	Type III Expected Mean Square
Environment (Env)	Var (Error) + 45.372 Var (IBlock (Env × Rep)) + Q (Env, Env × Genotypes)
Incomplete block (Env × block)	Var (Error) + 36.829 Var (IBlock (Env × Rep))
Accessions	Var (Error) + Q (Genotypes, Env × Genotypes)
Env × Accessions	Var (Error) + Q (Env × Genotypes)

Homogeneity and normality of variance were checked using Bartlett and Shapiro-Wilk statistics using R/package agricolae [47]; Least Square Means (Lsmeans) were estimated using R/package lsmeans [48]. Lsmeans were compared using Tukey's studentized range (HSD) (at p-value < 0.05). Pearson correlation analysis (r) was carried out between lsmeans using R/package corr.test [47]. Mean-based heritability (h^2) was estimated using the following model:

$$h^2 = \sigma_G^2 / [\sigma_G^2 + (\sigma_E^2 / ri)]$$

where σ_G^2 is the genetic variance, σ_E^2 the residual variance and ri is the number of replicates [49].

2.5. Association Mapping

The estimated Lsmeans for GPC and SNP markers were subjected to association analysis according to the following mixed linear model (MLM) in R package GAPIT [50].

$$Y = \mu + Zu + Wm + e \quad (1)$$

where Y is a vector of the total protein content, μ is a vector of intercepts, u is an n × 1 vector of random polygene background effects, e is a vector of random experimental errors with mean 0 and covariance matrix Var(e), Z is an incidence matrix relating Y to u. Var(u) = 2 KVg, where K is a known n × n matrix of a realized relationship matrix, estimated using the A.mat function in R software [51], as K = WW/C, where $W_{ik} = X_{ik} + 1 - 2_{pk}$ and $_{pk}$ is the frequency of the one allele at marker $_k$ [51], Vg is the unknown genetic variance, which is a scalar, m is a vector of fixed effects due to SNP markers, W is incidence matrix relating Y to m. Var(e) = RVR, where R is an n × n matrix, and VR is the unknown residual variance, which is a scalar too. Furthermore, principal component analysis (PCA) was conducted using the filtered SNP markers [52] and the integrated PCA function (prcomp) of the R software. In addition to Model (1), another three models were fitted. Model (2) contained the K matrix and the first PCA; Model (3) contained the K matrix, in addition to PCA1 and 2. Moreover, Model (4) contained the K matrix, in addition to the first three PCAs. p-values estimated from the mixed models were subjected to false discovery rate (FDR) corrections using Q-value estimates applied in the R package q-value [53]. The proportion of phenotypic variance explained (R^2) by the significant markers, and their additive effects were estimated using the GAPIT function, according to Wray et al. [54], in R software [50].

3. Results

3.1. Grain Protein Content (GPC)

Normal distribution and homogeneity of variance for grain protein content (GPC) were observed across the four environments (two seasons and two water regimes). Thus, combined analysis of variance across environments was conducted. Combined analysis of variance for GPC indicated a highly significant effect (p-value < 0.01) for the environments, genotypes, and genotype × environment interaction (Table 1). Broad-sense heritability estimates ranged from 0.49 to 0.60 for well-watered and water deficit conditions, respectively. Furthermore, the broad sense heritability estimates across years, and water regimes (the four environments) was 0.64 lsmeans of the grain protein content (GPC) ranged from 5.96 to 17.11% with a mean of 10.15 under well-watered conditions during 2016, and 6.88 to 17.43 with a mean of 9.67 in 2017 growing seasons. On the other hand, under water deficit conditions, GPC ranged from 11.12 to 18.5 with a mean of 14.9 in 2016 and 9.8 to 18.3 with a mean of 13.97 in 2017 growing seasons. Although, no significant difference was detected for the difference between means of the growing seasons, the difference between the lsmeans of the water regimes was highly significant, based on HSD at 0.01 probability level. Overall, our results indicated that water deficit increased GPC by 29% across the two growing seasons (Figure 1).

Table 1. Analysis of variance for grain protein content (GPC) of the 2111 genotypes across environments.

Source	DF	Type III SS	Mean Square	F Value
Environment	3	70,093.78	23,364.59	19,188.5 **
IBlock (Replicate Environment)	256	1361.56	5.31	4.37
Genotypes	2113	26,096.19	12.35	10.14 **
Environment × Genotypes	6255	26,164.38	4.18	3.44 **
Error	9208	11,211.99	1.21	

**: Significant at 0.01 probability level.

Furthermore, the correlation between GPC obtained from well-watered with that obtained water deficit across all genotypes was positive and significant ($r = 0.23$, p-value = 0.01). The first quartiles for the GPC across growing seasons (the cut off for the lowest 25%) under well-watered and water deficit conditions were ≤ 8.36 and 13.41, respectively (Figure 1). Whereas, the third quartile (the cut off for the highest 25%) of the genotypes under well-watered and water deficit conditions were ≥ 11.35 and 14.66, respectively. The first and third quartiles in this study were used as criteria to classify the genotypes into high and low GPC genotypes. Therefore genotypes with GPC ≤ 8.36 under well-watered and ≤ 13.41 under water deficit conditions, were defined as low protein genotypes. On the other hand, genotypes with GPC ≥ 11.35 under well-watered and ≥ 14.66 under water deficit conditions were defined as high protein genotypes. Grain protein content (GPC) for all genotypes under well-watered and water deficit conditions (Figure 2) indicated that 166 (7.8% of the genotypes) had high protein content under well-watered and water deficit conditions concurrently. Another, 200 genotypes (9.47%) were classified as low protein genotypes under both well-watered and water deficit conditions concurrently. The top 20 accessions with the highest GPC under well-watered and water deficit conditions are presented in Table 2, in which no overlapping accessions between the two water regimes were detected. Out of the top 20 accessions, detected under well-watered conditions, nine landraces were present. On the other hand, 18 landraces were present among the top 20 accessions detected under water deficit conditions. Overall, the estimated lsmeans from the landraces (882 accessions) under well-watered conditions was 10.9; which was 11.22% higher than the overall average of all other accessions (Table 2). Additionally, under water deficit conditions the average GPC for the landraces was 15.04 which was 7.9% higher than the overall average of all other accessions. Overall, our results indicate that moisture has a significant impact on GPC accumulation in wheat. Landraces had higher GPC, compared to other germplasm used in the current study.

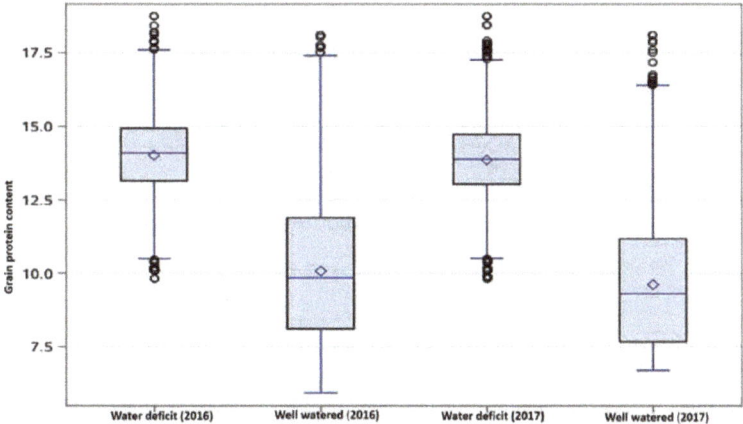

Figure 1. Boxplot for the overall performance of the 2111 wheat accessions across the four environments (well-watered and water-deficit conditions in 2016 and 2017 growing seasons).

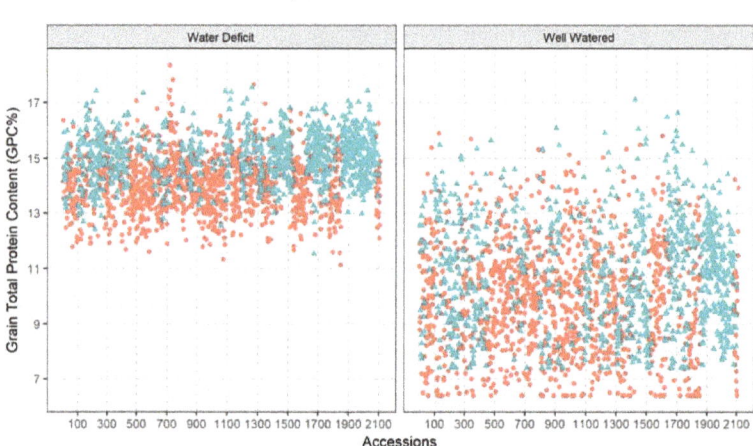

Figure 2. The overall performance of the 2111 wheat accessions across the two growing seasons under water deficit and well-watered growth conditions.

Table 2. Lsmean values of the grain protein content (GPC) of 20 accessions with the highest values across 2015/2016 and 2016/2017 growing seasons obtained from well-watered (control) and water-deficit conditions.

Well-Watered				Water Deficit			
Accession	Origin	Improvement	Mean	Accession	Origin	Improvement	Mean
534,406	Algeria	landrace	14.78	366,801	Afghanistan	landrace	17.965
534,448	Algeria	landrace	15.63	350,850	Austria	landrace	18.38
338,364	Belgium	cultivar	14.83	350,820	Austria	landrace	18.1125
481,731	Bhutan	landrace	14.69	565,254	Bolivia	landrace	17.9275
14,261	Canada	breeding	15.39	374,243	Chad	landrace	18.135
313,109	Colombia	uncertain	15.13	57,825	India	landrace	18.0175
372,434	Cyprus	landrace	14.90	382,048	Iran	landrace	18.535
428,672	Czech Republic	cultivar	15.33	625,916	Iran	landrace	18.43
254,023	Europe	uncertain	15.27	623,758	Iran	landrace	18.055
278,279	Greece	landrace	15.09	624,992	Iran	landrace	18.03
468,988	Greece	landrace	16.11	624,124	Iran	landrace	17.9125
15,396	Lebanon	uncertain	15.89	626,116	Iran	landrace	17.9075
520,369	Mexico	breeding	15.80	623,968	Iran	landrace	17.8525
525,283	Morocco	landrace	15.49	70,704	Iraq	landrace	18.42
477,901	Peru	landrace	15.05	191,987	Portugal	landrace	18.3475
370,724	Poland	cultivar	15.03	345,474	Serbia	landrace	18.3975
155,119	Russian Federation	cultivar	15.68	225,424	Uruguay	breeding	18.355
479,700	South Africa	cultivar	15.48	225,519	Uruguay	breeding	17.8375
241,596	Taiwan	cultivar	15.31	36,500	Uzbekistan	landrace	17.95
534,366	Tunisia	landrace	14.98	24,485	Uzbekistan	landrace	17.85

3.2. Association Mapping for Grain Protein Content

A total of 3215 mapped SNPs were used for estimating the extent of linkage disequilibrium (LD) in the 2111 wheat accessions. Only SNP loci having MAF ≥ 0.05 and missing values $\leq 10\%$ were used to estimate r^2 across all SNPs. The estimates of r^2 for all pairs of SNPs loci were used to determine the rate of LD decay with genetic distance. Across the three wheat genomes, i.e., A, B and D using only markers with significant r^2 (p-value = 0.001), the LD ranged from 0 to 0.35. Overall, LD declined to 50% of its initial value at about 8 cM (Supplementary Materials, Figure S1). Eigenvector decomposition of the kinship matrix was used to investigate the population structure among accessions. The first principal component (PCA) accounts for less than 1% of the total variance (Supplementary Materials, Figure S2). Nevertheless, GWAS models with kinship matrix (K matrix, supporting information Figure S3) with zero, one, two or three PCAs were compared using Bayesian information criteria (BIC). The results indicated noticeable difference between the four models. Additionally, the first model, i.e., with no PCA produced the highest BIC values, given that the largest is the best [55]. Therefore, we reported the results of association mapping using only the K matrix in which it accounted for most of the stratification among accessions.

Association mapping analysis was conducted on each environment separately (two growing seasons and two water regimes). Genome-wide association mapping (GWAS) indicated that 46 SNP markers found to be significantly linked with GPC. The significant SNP markers were located on chromosomes 1A (12 SNPs), 1B (12 SNPs), 1D (7 SNPs), 6A (6 SNPs), 6B (7 SNPs) and 6D (3 SNPs) (Figures 3 and 4). Out of the 46 significant SNP markers, ten markers were linked with GPC under well-watered and water deficit conditions in one growing season at least. Three SNP markers (IWA3169, IWA3501, and IWA7937) were significantly linked with GPC across the four environments (2016, 2017 growing seasons, and well-watered and water deficit conditions) (Table 3). Four markers (IWA6649, IWA6787, IWA3481 and IWA4351) found to be linked with GPC in three environments (2016 well-watered, 2016 and 2017 water deficit conditions) (Table 3). These results together indicate that some loci were significantly associated with GPC in wheat irrespective of water status.

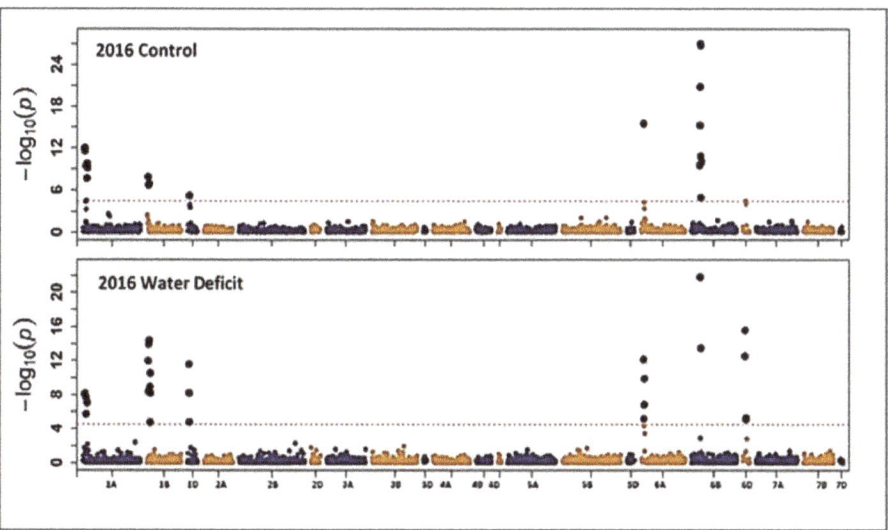

Figure 3. Manhattan plot for grain protein content (GPC) obtained from genome-wide association mapping in the 2016 growing season.

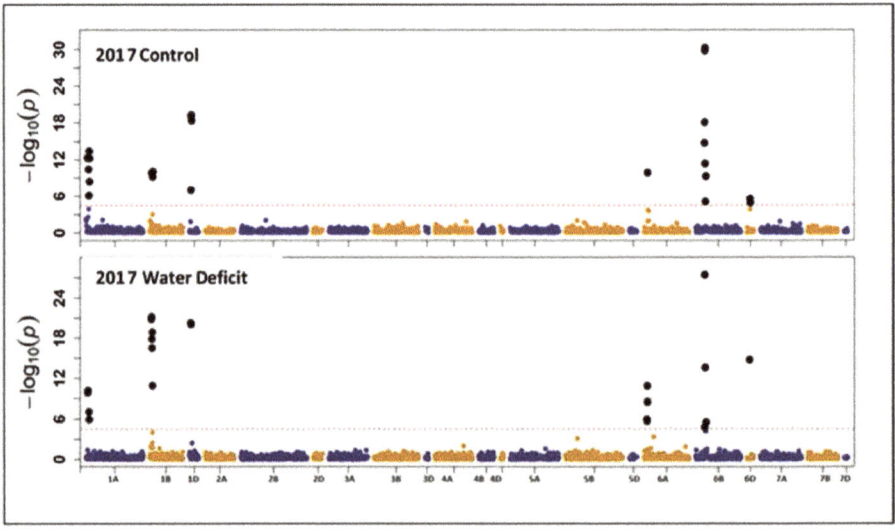

Figure 4. Manhattan plot for grain protein content (GPC) obtained from genome-wide association mapping in the 2017 growing season.

Table 3. SNP markers that found to be significantly linked with GPC under well-watered (control) and water deficit conditions.

Marker	Chrom	Position	Well-Watered 2016	Well-Watered 2017	Water Deficit 2016	Water Deficit 2017	R^2 (%)	Additive Effect	MAF	Marker	Chrom	Position	Well-Watered 2016	Well-Watered 2017	Water Deficit 2016	Water Deficit 2017	R^2 (%)	Additive Effect	MAF
IWA5150	1A	9.9	+	+	−	−	0.893	−0.007	0.19	IWA8551	1D	32.8	−	−	+	−	1.069	0.062	0.25
IWA6649	1A	11.6	+	+	+	+	1.141	0.07	0.35	IWA3481	1D	45.1	−	+	+	+	1.122	0.13	0.07
IWA4351	1A	11.6	+	−	−	−	1.089	0.062	0.35	IWA3446	1D	45.1	−	+	+	+	1.001	0.086	0.07
IWA4643	1A	21	+	+	−	−	0.892	0.004	0.28	IWA5020	1D	47.7	−	+	−	−	0.918	−0.026	0.33
IWA4753	1A	21.7	−	+	−	−	0.9	−0.02	0.1	IWA5019	1D	47.7	−	+	−	−	0.918	−0.026	0.33
IWA7191	1A	21.7	−	+	−	+	0.965	0.053	0.13	IWA5018	1D	47.7	−	+	−	−	0.917	−0.026	0.33
IWA4678	1A	22.5	−	+	−	−	0.901	0.028	0.08	IWA4598	1D	48.6	−	+	−	−	0.912	0.023	0.33
IWA4644	1A	22.9	−	+	−	−	0.915	−0.026	0.16	IWA7007	6A	10	−	−	+	−	0.918	0.025	0.23
IWA4754	1A	23.2	−	+	−	−	0.892	0.003	0.34	IWA4551	6A	16.2	+	−	−	−	1.294	−0.13	0.13
IWA4506	1A	26.9	−	−	−	−	0.907	−0.019	0.29	IWA4552	6A	16.2	+	−	−	−	1.313	−0.131	0.13
IWA7050	1A	32.5	−	+	+	+	1.305	0.092	0.38	IWA7288	6A	17.8	−	−	+	+	1.316	0.093	0.35
IWA4163	1A	32.8	+	−	−	−	1.17	−0.076	0.39	IWA7287	6A	21.9	−	−	+	+	1.391	0.116	0.21
IWA4349	1B	13.2	+	−	−	−	1.589	0.128	0.26	IWA4962	6A	22.8	−	+	−	−	0.923	−0.027	0.24
IWA6787	1B	13.2	+	−	+	−	1.433	0.105	0.28	IWA4730	6A	48.5	+	+	+	+	0.922	0.046	0.06
IWA7048	1B	22.9	−	−	+	+	1.746	0.244	0.07	IWA3501	6B	48.8	+	+	+	+	2.681	0.213	0.39
IWA7480	1B	22.9	−	−	−	+	1.472	0.111	0.35	IWA7937	6B	48.8	+	+	+	+	2.654	0.187	0.37
IWA3169	1B	23.7	+	+	−	+	2.027	0.161	0.31	IWA3923	6B	48.8	+	+	+	−	1.208	0.117	0.11
IWA8199	1B	27.4	+	−	+	+	1.271	0.099	0.25	IWA6466	6B	48.8	+	+	+	−	1.448	0.131	0.18
IWA7345	1B	28.1	−	−	−	+	1.808	0.23	0.09	IWA6467	6B	48.8	+	+	+	−	1.455	0.132	0.18
IWA6611	1B	28.1	−	−	−	−	1.636	0.131	0.27	IWA5986	6B	50.8	+	+	−	−	0.892	0.005	0.24
IWA6610	1B	28.1	−	−	+	+	1.642	−0.13	0.27	IWA6673	6D	17.2	−	−	+	−	1.147	0.079	0.24
IWA3738	1B	28.2	−	+	−	−	1.451	0.12	0.22	IWA3624	6D	17.3	−	−	−	−	1.073	0.068	0.41
IWA8275	1B	28.2	−	−	+	+	1.618	0.128	0.27	IWA7616	6D	29.8	−	−	−	+	1.542	0.142	0.17

— and + refer to nonsignificant and significant SNPs, respectively.

Under well-watered conditions for the two growing seasons, seven SNP markers (IWA5150, IWA4643, IWA4754, IWA3923, IWA6466, IWA6467, and IWA5986) found to be significantly linked with GPC. On the other hand, under water deficit conditions for the two growing seasons, six SNP markers (IWA7191, IWA8199, IWA7345, IWA3446, IWA7288, and IWA7287) found to be significantly linked with GPC. In contrary, 14 markers (IWA4753, IWA4678, IWA4644, IWA4506, IWA4163, IWA3738, IWA5020, IWA5019, IWA5018, IWA4598, IWA4551, IWA4552, IWA4962, and IWA4730) found to be significantly linked with GPC during only one growing season under well-watered conditions. Another ten markers (IWA7616, IWA3624, IWA6673, IWA7007, IWA8551, IWA6610, IWA6611, IWA7480, IWA7048, and IWA7050) found to be significantly linked with GPC under water deficit conditions in only one growing season. Repeatability of the GPC associated loci in 2 seasons under any given water treatment suggests the feasibility of using/developing markers in LD with these loci.

4. Discussion

Protein content is an essential compositional trait in wheat, which has a broad impact in the food industry concerning human nutrition and health. Consequently, breeding for enhanced end-use quality is one of the essential breeding goals in wheat. However, GPC in wheat is positively affected by water deficit compared to well-watered conditions [10]. In this study, we seek to evaluate a comprehensive spring wheat collection for grain protein content (GPC) and to locate genomic regions associated with GPC under well-watered and water deficit conditions using GWAS approach.

The most striking observation in this study was the weak, positive and significant correlation between GPC obtained from the well-watered condition and water deficit conditions ($r = 0.23$). That weak correlation implies strong genotype × environment interaction, in which genotypes responded differently concerning water treatment. Increase in GPC under water deficit conditions could be mainly due to higher rates of accumulation of grain nitrogen and lower rates of accumulation of carbohydrates. High moisture, on the other hand, may decrease GPC by dilution of nitrogen with carbohydrates [56]. An increased grain protein and gluten content in response to water deficit as compared to the well-watered experiment in a winter wheat was also reported in a previous study [57]. The current study, as well as previous reports, indicated a significant effect of environment (moisture and growing seasons) on wheat GPC accumulation. Analysis of variance indicated a significant effect of moisture, genotype, and genotype × environment interaction on GPC in wheat, suggesting that GPC is a complex trait influenced by several factors. The significant genotypic effect observed in this study also indicated a wide range of variation for GPC accumulation among wheat accessions used. Moreover, around 366 (166 with high GPC and 200 with low GPC) wheat genotypes performed relatively the same across environments, which implies that GPC accumulation on these genotypes was less responsive to moisture.

Genotypic variation is a result of several alleles on genes which result in different responses to environmental conditions [58]. Furthermore, landraces serve as a valuable genetic resource in which it might provide new alleles for improvement of economically important traits such as GPC [19]. Results reported herein showed that landraces outperformed cultivated genotypes concerning GPC. These findings agree with previous reports [59,60] in which 121 landraces, 101 obsolete cultivars, and modern wheat cultivars were evaluated for GPC under the same environmental conditions, and landraces had higher total protein content compared to other studied accessions. Grain quality of some wheat landraces should be of particular interest because much broader diversity can be found in landraces compared to modern wheat cultivars [61]. Additionally, most of the organic wheat production systems rely on cultivars that were developed for high-input production systems [60,62]. Wheat landraces have been developed mostly in environments with low nutrient availability; they represent a source of variation for selection of genotypes adapted to cropping systems with low fertilizer input [61]. In the current study, we identified 224, 214 and 70 wheat landraces that were found to have high GPC under well-watered, water deficit and both conditions, respectively. Our results and previous reports indicated that GPC depends mainly on genotype, environment, and genotype × environment interaction [59]. However, the response mechanism that modifies protein accumulation under water deficit conditions is still unclear. Recently, a putative mechanism underlying the increased accumulation of storage proteins in wheat endosperm under water deficit was provided by Chen et al. [63]. They identified four differentially expressed miRNAs induced by drought stress that may affect the development of protein bodies in caryopsis by regulating the expression levels of target genes involved in protein biosynthesis pathways.

One of the primary goals of this study was to locate significant genomic regions that control the accumulation of GPC which might shed light on the genetic architecture of GPC and the protein accumulation mechanism. The genome-wide association mapping analysis, applied in the current study, using the kinship (K) matrix in a mixed model indicated that K matrix was adequate in accounting for population structure [64]. Also, these results agree with those of Zhao et al. [65], in which they found that K models were adequate for genome-wide association mapping. Furthermore, the K model was more effective in reducing the false-positive rate compared to using the Q + K model. Linkage disequilibrium (LD) was estimated using r^2 among all pairs of SNPs loci, in which r^2 in this study was 0.09, which is higher than that obtained by Breseghello and Sorrells [66] and 0.019 reported by Neumann et al. [67] because of their small size populations, and with a similar number of marker pairs. This indicates that the population size might have an impact on the LD.

Genome-wide association analysis (GWAS) was conducted on each environment separately to measure the repeatability of the significant SNPs, and the effect of moisture on the genomic regions controlling GPC. Several SNPs found to be significantly linked to the GPC under well-watered conditions but not significantly linked to GPC under water deficit conditions and *vice versa*. Moreover, ten QTLs were linked with GPC under both well-watered and water deficit conditions. The GWAS analysis suggested a significant role of genotype × environment interaction in detecting GPC associated loci. Genome-wide association studies using diverse wheat germplasm have successfully detected GPC associated loci in durum wheat [68], and bread wheat lines [69]. Thus, the SNPs associated with GPC under water deficit or well-watered environmental conditions, from this study might provide useful molecular information for wheat breeders to incorporate specific QTLs to increase GPC in low input or drought-stressed environments. Around 50% of the significant SNPs detected in the current study was on chromosome 1, where copies of *Glu-B1* and *Gli-B1* genes reside [70]. *Glu-B1* and *Gli-B1* genes were previously reported to contribute of about 24.6 and 19.5% of the total phenotypic variation for sedimentation volume (determines gluten strength and in turn cooking quality of pasta) [2]. Several SNP loci in LD with sedimentation volume were discovered recently on chromosome 1A and 1B, in durum wheat [68].

These results together emphasized the importance of using diverse worldwide germplasm to dissect the genetic architecture of GPC in wheat and identify accessions that might be potential parents in wheat breeding programs. Ongoing multiple years, multiple replication study using 406 accessions identified in the current study is being conducted, to evaluate these genotypes for yield and validate the GPC associated loci detected herein. Furthermore, GPC estimates under well-watered and water deficit conditions was used as a selection parameter to downsize the number of accessions from 2111 to 406. Reducing the number of accessions will allow us to profoundly investigate other wheat quality aspects such as concentrations (soluble and insoluble) of glutenin, α/β, γ gliadin and albumin/globulin in addition to the total protein for high and low GPC genotypes.

5. Conclusions

Based on previous research and our findings, the spring wheat collection used in this study contains high protein accessions. Furthermore, GPC measurement under well-watered and water deficit conditions was used as a selection criterion to reduce the number of accessions from 2111 to 406 accessions. This reduction in the number of studied accessions will allow us to profoundly study other wheat quality aspects such as concentrations (soluble and insoluble) of glutenin, α/β, γ gliadin and albumin/globulin in addition to the total protein for high and low GPC genotypes. It also represents a precious resource for further investigations including annotation of relevant genomic regions/genes using available wheat genomic resources to study the GPC. Results of GWAS indicated that several genomic regions were involved in GPC accumulation in wheat grains. Furthermore, GWAS results also suggested a significant role for genotype x environment interaction in the identification of GPC associated loci under well-watered and water deficit conditions. The identified loci might allow development of marker-assisted selection (MAS) for GPC and might also facilitate the development of a better understanding of the genetic architecture that controls GPC in wheat. Therefore, the high and low GPC accessions identified in the current study were included in ongoing multiple years and locations studies to evaluate them for yield and confirm the GPC associated loci detected.

Supplementary Materials: The following are available online at http://www.mdpi.com/2223-7747/7/3/56/s1, Figure S1: Decay of r^2 as a function of genetic distance between SNP markers estimated for 2111 spring wheat collection from different geographic regions, Figure S2: The percentage of variance explained by principal components (PCA), Figure S3: Heatmap and dendrogram of a kinship matrix estimated using the A.mat function (rrBLUP package) based on 5090 SNPs among 2111 wheat accessions.

Author Contributions: I.S.E. Conception or design of the work, collecting the phenotypic data, data analysis, drafting the article, final approval of the version to be published. S.A.M. Design of the work, collecting the phenotypic data, review of the first draft, final approval of the version to be published. R.K.R. Design of the work, review of the first draft, final approval of the version to be published. A.M.K.N. Review of the first draft, final approval of the version to be published.

Funding: This study was supported financially by the Science and Technology Development Fund (STDF), Egypt, Grant No. 14935.

Acknowledgments: The authors would like to thank Holland Computing Center (University of Nebraska-Lincoln) for allowing the authors of this publication to use UNL's supercomputing resources.

Conflicts of Interest: The authors declare no conflict of interest.

References

1. Hawkesford, M.J. Reducing the reliance on nitrogen fertilizer for wheat production. *J. Cereal Sci.* **2014**, *59*, 276–283. [CrossRef] [PubMed]
2. Würschum, T.; Leiser, W.L.; Kazman, E.; Longin, C.F.H. Genetic control of protein content and sedimentation volume in European winter wheat cultivars. *Theor. Appl. Genet.* **2016**, *129*, 1685–1696. [CrossRef] [PubMed]
3. Day, L. Proteins from land plants? Potential resources for human nutrition and food security. *Trends Food Sci. Technol.* **2013**, *32*, 25–42. [CrossRef]
4. Cantu, D.; Pearce, S.P.; Distelfeld, A.; Christiansen, M.W.; Uauy, C.; Akhunov, E.; Fahima, T.; Dubcovsky, J. Effect of the down-regulation of the high Grain Protein Content (GPC) genes on the wheat transcriptome during monocarpic senescence. *BMC Genome* **2011**, *12*, 492. [CrossRef] [PubMed]
5. Triboi, E. Environmentally-induced changes in protein composition in developing grains of wheat are related to changes in total protein content. *J. Exp. Bot.* **2003**, *54*, 1731–1742. [CrossRef] [PubMed]
6. Bertheloot, J.; Martre, P.; Andrieu, B. Dynamics of Light and Nitrogen Distribution during Grain Filling within Wheat Canopy. *Plant Physiol.* **2008**, *148*, 1707–1720. [CrossRef] [PubMed]
7. Beta, T.; Nam, S.; Dexter, J.E.; Sapirstein, H.D. Phenolic Content and Antioxidant Activity of Pearled Wheat and Roller-Milled Fractions. *Cereal Chem. J.* **2005**, *82*, 390–393. [CrossRef]
8. Li, X.; Zhou, L.; Liu, F.; Zhou, Q.; Cai, J.; Wang, X.; Dai, T.; Cao, W.; Jiang, D. Variations in Protein Concentration and Nitrogen Sources in Different Positions of Grain in Wheat. *Front. Plant Sci.* **2016**, *7*, 942. [CrossRef] [PubMed]
9. Arduini, I.; Masoni, A.; Ercoli, L.; Mariotti, M. Grain yield, and dry matter and nitrogen accumulation and remobilization in durum wheat as affected by variety and seeding rate. *Eur. J. Agron.* **2006**, *25*, 309–318. [CrossRef]
10. Lantican, M.A.; Braun, H.J.; Payne, T.S.; Singh, R.P.; Sonder, K.; Baum, M.; van Ginkel, M.; Erenstein, O. *Impacts of International Wheat Improvement Research, 1994–2014*; CIMMYT: Texcoco de Mora, Mexico, 2016; ISBN 9786078263554.
11. Ashraf, M. Stress-Induced Changes in Wheat Grain Composition and Quality. *Crit. Rev. Food Sci. Nutr.* **2014**, *54*, 1576–1583. [CrossRef] [PubMed]

12. Blanco, A.; Mangini, G.; Giancaspro, A.; Giove, S.; Colasuonno, P.; Simeone, R.; Signorile, A.; De Vita, P.; Mastrangelo, A.M.; Cattivelli, L.; et al. Relationships between grain protein content and grain yield components through quantitative trait locus analyses in a recombinant inbred line population derived from two elite durum wheat cultivars. *Mol. Breed* **2012**, *30*, 79–92. [CrossRef]
13. De Santis, M.A.; Giuliani, M.M.; Giuzio, L.; De Vita, P.; Lovegrove, A.; Shewry, P.R.; Flagella, Z. Differences in gluten protein composition between old and modern durum wheat genotypes in relation to 20th century breeding in Italy. *Eur. J. Agron.* **2017**, *87*, 19–29. [CrossRef] [PubMed]
14. Ravier, C.; Meynard, J.M.; Cohan, J.P.; Gate, P.; Jeuffroy, M.H. Early nitrogen deficiencies favor high yield, grain protein content and N use efficiency in wheat. *Eur. J. Agron.* **2017**, *89*, 16–24. [CrossRef]
15. Gaju, O.; Allard, V.; Martre, P.; Le Gouis, J.; Moreau, D.; Bogard, M.; Hubbart, S.; Foulkes, M.J. Nitrogen partitioning and remobilization in relation to leaf senescence, grain yield and grain nitrogen concentration in wheat cultivars. *Field Crops Res.* **2014**, *155*, 213–223. [CrossRef]
16. Mondal, S.; Rutkoski, J.E.; Velu, G.; Singh, P.K.; Crespo-Herrera, L.A.; Guzmán, C.; Bhavani, S.; Lan, C.; He, X.; Singh, R.P. Harnessing Diversity in Wheat to Enhance Grain Yield, Climate Resilience, Disease and Insect Pest Resistance and Nutrition Through Conventional and Modern Breeding Approaches. *Front. Plant Sci.* **2016**, *7*, 991. [CrossRef] [PubMed]
17. Shewry, P.R.; Hey, S.J. The contribution of wheat to human diet and health. *Food Energy Secur.* **2015**, *4*, 178–202. [CrossRef] [PubMed]
18. Lindeque, R.C. Protein Quality vs. Quantity in South African Commercial Bread Wheat Cultivars. Ph.D. Thesis, University of the Free State, Bloemfontein, South Africa, 2016.
19. Soriano, J.M.; Villegas, D.; Aranzana, M.J.; García Del Moral, L.F.; Royo, C. Genetic structure of modern durum wheat cultivars and mediterranean landraces matches with their agronomic performance. *PLoS ONE* **2016**, *11*, e0160983. [CrossRef] [PubMed]
20. Shewry, P.R. Improving the protein content and composition of cereal grain. *J. Cereal Sci.* **2007**, *46*, 239–250. [CrossRef]
21. Velu, G.; Singh, R.P.; Cardenas, M.E.; Wu, B.; Guzman, C.; Ortiz-Monasterio, I. Characterization of grain protein content gene (GPC-B1) introgression lines and its potential use in breeding for enhanced grain zinc and iron concentration in spring wheat. *Acta Physiol. Plant.* **2017**, *39*, 212. [CrossRef]
22. Uauy, C.; Brevis, J.C.; Dubcovsky, J. The high grain protein content gene Gpc-B1 accelerates senescence and has pleiotropic effects on protein content in wheat. *J. Exp. Bot.* **2006**, *57*, 2785–2794. [CrossRef] [PubMed]
23. Amiri, R.; Bahraminejad, S.; Sasani, S.; Jalali-Honarmand, S.; Fakhri, R. Bread wheat genetic variation for grain's protein, iron and zinc concentrations as uptake by their genetic ability. *Eur. J. Agron.* **2015**, *67*, 20–26. [CrossRef]
24. Ravel, C.; Praud, S.; Murigneux, A.; Linossier, L.; Dardevet, M.; Balfourier, F.; Dufour, P.; Brunel, D.; Charmet, G. Identification of Glu-B1-1 as a candidate gene for the quantity of high-molecular-weight glutenin in bread wheat (*Triticum aestivum* L.) by means of an association study. *Theor. Appl. Genet.* **2006**, *112*, 738–743. [CrossRef] [PubMed]
25. Nakamura, T.; Yamamori, M.; Hirano, H.; Hidaka, S. Decrease of Waxy (Wx) Protein in Two Common Wheat Cultivars with Low Amylose Content. *Plant Breed* **1993**, *111*, 99–105. [CrossRef]
26. Giroux, M.J.; Morris, C.F. A glycine to serine change in puroindoline b is associated with wheat grain hardness and low levels of starch-surface friabilin. *TAG Theor. Appl. Genet.* **1997**, *95*, 857–864. [CrossRef]
27. Gupta, R.B.; Singh, N.K.; Shepherd, K.W. The cumulative effect of allelic variation in LMW and HMW glutenin subunits on dough properties in the progeny of two bread wheats. *Theor. Appl. Genet.* **1989**, *77*, 57–64. [CrossRef] [PubMed]
28. Araki, E.; Miura, H.; Sawada, S. Identification of genetic loci affecting amylose content and agronomic traits on chromosome 4A of wheat. *TAG Theor. Appl. Genet.* **1999**, *98*, 977–984. [CrossRef]
29. Payne, P.I.; Nightingale, M.A.; Krattiger, A.F.; Holt, L.M. The relationship between HMW glutenin subunit composition and the bread-making quality of British-grown wheat varieties. *J. Sci. Food Agric.* **1987**, *40*, 51–65. [CrossRef]
30. McCartney, C.A.; Somers, D.J.; Lukow, O.; Ames, N.; Noll, J.; Cloutier, S.; Humphreys, D.G.; McCallum, B.D. QTL analysis of quality traits in the spring wheat cross RL4452 × 'AC Domain'. *Plant Breed* **2006**, *125*, 565–575. [CrossRef]

31. Sun, H.; Lü, J.; Fan, Y.; Zhao, Y.; Kong, F.; Li, R.; Wang, H.; Li, S. Quantitative trait loci (QTLs) for quality traits related to protein and starch in wheat. *Prog. Nat. Sci.* **2008**, *18*, 825–831. [CrossRef]
32. Fu, Y.-B.; Yang, M.-H.; Zeng, F.; Biligetu, B. Searching for an Accurate Marker-Based Prediction of an Individual Quantitative Trait in Molecular Plant Breeding. *Front. Plant Sci.* **2017**, *8*, 1182. [CrossRef] [PubMed]
33. Zhao, Y.; Mette, M.F.; Gowda, M.; Longin, C.F.H.; Reif, J.C. Bridging the gap between marker-assisted and genomic selection of heading time and plant height in hybrid wheat. *Heredity* **2014**, *112*, 638–645. [CrossRef] [PubMed]
34. Adhikari, T.B.; Gurung, S.; Hansen, J.M.; Jackson, E.W.; Bonman, J.M. Association Mapping of Quantitative Trait Loci in Spring Wheat Landraces Conferring Resistance to Bacterial Leaf Streak and Spot Blotch. *Plant Genome J.* **2012**, *5*, 1–16. [CrossRef]
35. Odong, T.L.; van Heerwaarden, J.; Jansen, J.; van Hintum, T.J.L.; van Eeuwijk, F.A. Determination of genetic structure of germplasm collections: Are traditional hierarchical clustering methods appropriate for molecular marker data? *Theor. Appl. Genet.* **2011**, *123*, 195–205. [CrossRef] [PubMed]
36. Zhang, L.; Liu, D.; Guo, X.; Yang, W.; Sun, J.; Wang, D.; Sourdille, P.; Zhang, A. Investigation of genetic diversity and population structure of common wheat cultivars in northern China using DArT markers. *BMC Genet.* **2011**, *12*, 42. [CrossRef] [PubMed]
37. Hiremath, P.J.; Kumar, A.; Penmetsa, R.V.; Farmer, A.; Schlueter, J.A.; Chamarthi, S.K.; Whaley, A.M.; Carrasquilla-Garcia, N.; Gaur, P.M.; Upadhyaya, H.D.; et al. Large-scale development of cost-effective SNP marker assays for diversity assessment and genetic mapping in chickpea and comparative mapping in legumes. *Plant Biotechnol. J.* **2012**, *10*, 716–732. [CrossRef] [PubMed]
38. Tadesse, W.; Ogbonnaya, F.C.; Jighly, A.; Sanchez-Garcia, M.; Sohail, Q.; Rajaram, S.; Baum, M. Genome-wide association mapping of yield and grain quality traits in winter wheat genotypes. *PLoS ONE* **2015**, *10*, e0141339. [CrossRef] [PubMed]
39. Uauy, C. Wheat genomics comes of age. *Curr. Opin. Plant Biol.* **2017**, *36*, 142–148. [CrossRef] [PubMed]
40. Taulemesse, F.; Gouis, J. Le; Gouache, D.; Gibon, Y.; Allard, V. Bread wheat (*Triticum aestivum* L.) grain protein concentration is related to early post-flowering nitrate uptake under putative control of plant satiety level. *PLoS ONE* **2016**, *11*, e0149668. [CrossRef] [PubMed]
41. Federer, W.T.; Crossa, J. I.4 Screening Experimental Designs for Quantitative Trait Loci, Association Mapping, Genotype-by Environment Interaction, and Other Investigations. *Front. Physiol.* **2012**, *3*, 156. [CrossRef] [PubMed]
42. Cavanagh, C.R.; Chao, S.; Wang, S.; Huang, B.E.; Stephen, S.; Kiani, S.; Forrest, K.; Saintenac, C.; Brown-Guedira, G.L.; Akhunova, A.; et al. Genome-wide comparative diversity uncovers multiple targets of selection for improvement in hexaploid wheat landraces and cultivars. *Proc. Natl. Acad. Sci. USA* **2013**, *110*, 8057–8062. [CrossRef] [PubMed]
43. Breiman, L. Random forests. *Mach. Learn.* **2001**, *45*, 5–32. [CrossRef]
44. Stekhoven, D.J.; Bühlmann, P. MissForest-non-parametric missing value imputation for mixed-type data. *Bioinformatics* **2012**, *28*, 112–118. [CrossRef] [PubMed]
45. Wang, S.; Wong, D.; Forrest, K.; Allen, A.; Chao, S.; Huang, B.E.; Maccaferri, M.; Salvi, S.; Milner, S.G.; Cattivelli, L.; et al. Characterization of polyploid wheat genomic diversity using a high-density 90,000 single nucleotide polymorphism array. *Plant Biotechnol. J.* **2014**, *12*, 787–796. [CrossRef] [PubMed]
46. Federer, W.T.; King, F. *Variations on Split Plot and Split Block Experiment Designs*; John Wiley & Sons: New York, NY, USA, 2006; ISBN 9780470108581.
47. R Core Team. *R: A Language and Environment for Statistical Computing*; R Foundation for Statistical Computing: Vienna, Austria, 2017; Available online: https://www.R-project.org/ (accessed on 6 October 2017).
48. Lenth, R.V. Least-Squares Means: The R Package lsmeans. *J. Stat. Softw.* **2016**, *69*. [CrossRef]
49. Asadabadi, Y.Z.; Khodarahmi, M.; Nazeri, S.M.; Peyghambari, S.A. Genetic Study of Grain Yield and its Components in Bread Wheat Using Generation Mean Analysis under Water Stress Condition. *J. Plant Physiol. Breed* **2012**, *2*, 55–60.
50. Lipka, A.E.; Tian, F.; Wang, Q.; Peiffer, J.; Li, M.; Bradbury, P.J.; Gore, M.A.; Buckler, E.S.; Zhang, Z. GAPIT: Genome association and prediction integrated tool. *Bioinformatics* **2012**, *28*, 2397–2399. [CrossRef] [PubMed]
51. Endelman, J.B.; Jannink, J.-L. Shrinkage Estimation of the Realized Relationship Matrix. *Genes Genome Genet.* **2012**, *2*, 1405–1413. [CrossRef] [PubMed]

52. Zheng, X.; Weir, B.S. Eigenanalysis of SNP data with an identity by descent interpretation. *Theor. Popul. Biol.* **2016**, *107*, 65–76. [CrossRef] [PubMed]
53. Storey, J.D.; Tibshirani, R. Statistical significance for genomewide studies. *Proc. Natl. Acad. Sci. USA* **2003**, *100*, 9440–9445. [CrossRef] [PubMed]
54. Wray, N.R.; Yang, J.; Hayes, B.J.; Price, A.L.; Goddard, M.E.; Visscher, P.M. Pitfalls of predicting complex traits from SNPs. *Nat. Rev. Genet.* **2013**, *14*, 507–515. [CrossRef] [PubMed]
55. Schwarz, G. Estimating the Dimension of a Model. *Ann. Stat.* **1978**, *6*, 461–464. [CrossRef]
56. Singh, S.; Gupta, A.K.; Kaur, N. Influence of drought and sowing time on protein composition, antinutrients, and mineral contents of wheat. *Sci. World J.* **2012**, *2012*, 485751. [CrossRef] [PubMed]
57. Ozturk, A.; Aydin, F. Effect of water stress at various growth stages on some quality characteristics of winter wheat. *J. Agron. Crop Sci.* **2004**, *190*, 93–99. [CrossRef]
58. Grenier, S.; Barre, P.; Litrico, I. Phenotypic Plasticity and Selection: Nonexclusive Mechanisms of Adaptation. *Scientifica* **2016**, *2016*, 7021701. [CrossRef] [PubMed]
59. Dvořáček, V.; Čurn, V. Evaluation of protein fractions as biochemical markers for identification of spelt wheat cultivars (*Triticum spelta* L.). *Plant Soil Environ.* **2003**, *49*, 99–105. [CrossRef]
60. Jaradat, A.A. Wheat landraces: A mini review. *Emir. J. Food Agric.* **2013**, *25*, 20–29. [CrossRef]
61. Lopes, M.S.; El-Basyoni, I.; Baenziger, P.S.; Singh, S.; Royo, C.; Ozbek, K.; Aktas, H.; Ozer, E.; Ozdemir, F.; Manickavelu, A.; et al. Exploiting genetic diversity from landraces in wheat breeding for adaptation to climate change. *J. Exp. Bot.* **2015**, *66*, 3477–3486. [CrossRef] [PubMed]
62. Baenziger, P.S.; Salah, I.; Little, R.S.; Santra, D.K.; Regassa, T.; Wang, M.Y. Structuring an Efficient Organic Wheat Breeding Program. *Sustainability* **2011**, *3*, 1190–1205. [CrossRef]
63. Chen, X.-Y.; Yang, Y.; Ran, L.-P.; Dong, Z.; Zhang, E.-J.; Yu, X.-R.; Xiong, F. Novel Insights into miRNA Regulation of Storage Protein Biosynthesis during Wheat Caryopsis Development under Drought Stress. *Front. Plant Sci.* **2017**, *8*, 1707. [CrossRef] [PubMed]
64. Endelman, J. Using rrBLUP 4.0. Jeffrey Endelman 17 September 2012. Available online: http://www.afhalifax.ca/bete/revive-eponge/Rblkfct/vignette.pdf (accessed on 8 July 2018).
65. Zhao, K.; Aranzana, M.J.; Kim, S.; Lister, C.; Shindo, C.; Tang, C.; Toomajian, C.; Zheng, H.; Dean, C.; Marjoram, P.; et al. An Arabidopsis example of association mapping in structured samples. *PLoS Genet.* **2007**, *3*, e4. [CrossRef] [PubMed]
66. Breseghello, F.; Sorrells, M.M.E. Association mapping of kernel size and milling quality in wheat (*Triticum aestivum* L.) cultivars. *Genetics* **2006**, *172*, 1165–1177. [CrossRef] [PubMed]
67. Neumann, K.; Kobiljski, B.; Denčić, S.; Varshney, R.K.; B?rner, A. Genome-wide association mapping: A case study in bread wheat (*Triticum aestivum* L.). *Mol. Breed* **2011**, *27*, 37–58. [CrossRef]
68. Fiedler, J.D.; Salsman, E.; Liu, Y.; Michalak de Jiménez, M.; Hegstad, J.B.; Chen, B.; Manthey, F.A.; Chao, S.; Xu, S.; Elias, E.M.; et al. Genome-Wide Association and Prediction of Grain and Semolina Quality Traits in Durum Wheat Breeding Populations. *Plant Genome* **2017**, *10*. [CrossRef] [PubMed]
69. Liu, J.; Feng, B.; Xu, Z.; Fan, X.; Jiang, F.; Jin, X.; Cao, J.; Wang, F.; Liu, Q.; Yang, L.; et al. A genome-wide association study of wheat yield and quality-related traits in southwest China. *Mol. Breed* **2018**, *38*, 1. [CrossRef]
70. Troccoli, A.; Borrelli, G.M.; De Vita, P.; Fares, C.; Di Fonzo, N. Mini Review: Durum Wheat Quality: A Multidisciplinary Concept. *J. Cereal Sci.* **2000**, *32*, 99–113. [CrossRef]

© 2018 by the authors. Licensee MDPI, Basel, Switzerland. This article is an open access article distributed under the terms and conditions of the Creative Commons Attribution (CC BY) license (http://creativecommons.org/licenses/by/4.0/).

Article

Effect of Selenium on the Responses Induced by Heat Stress in Plant Cell Cultures

Massimo Malerba [1] and Raffaella Cerana [2,*]

[1] Dipartimento di Biotecnologie e Bioscienze, Università degli Studi di Milano-Bicocca, 20126 Milan, Italy; massimo.malerba@unimib.it
[2] Dipartimento di Scienze dell'Ambiente e della Terra, Università degli Studi di Milano-Bicocca, 20126 Milan, Italy
* Correspondence: raffaella.cerana@unimib.it; Tel.: +39-02-64482932

Received: 11 July 2018; Accepted: 10 August 2018; Published: 11 August 2018

Abstract: High temperatures are a significant stress factor for plants. In fact, many biochemical reactions involved in growth and development are sensitive to temperature. In particular, heat stress (HS) represents a severe issue for plant productivity and strategies to obtain high yields under this condition are important goals in agriculture. While selenium (Se) is a nutrient for humans and animals, its role as a plant micronutrient is still questioned. Se can prevent several abiotic stresses (drought, heat, UV, salinity, heavy metals), but the action mechanisms are poorly understood. Se seems to regulate reactive oxygen species (ROS) and to inhibit heavy metals transport. In addition, it has been demonstrated that Se is essential for a correct integrity of cell membranes and chloroplasts, especially the photosynthetic apparatus. Previous results showed that in tobacco (*Nicotiana tabacum* cv. Bright-Yellow 2) cultures HS (5 min at 50 °C) induced cell death with apoptotic features, accompanied by oxidative stress and changes in the levels of stress-related proteins. In this work we investigated the effect of Se on the responses induced by HS. The obtained results show that Se markedly reduces the effects of HS on cell vitality, cytoplasmic shrinkage, superoxide anion production, membrane lipids peroxidation, activity of caspase-3-like proteases, and the levels of some stress-related proteins (Hsp90, BiP, 14-3-3s, cytochrome *c*).

Keywords: cell death; heat stress; plant cell cultures; selenium; tobacco BY-2

1. Introduction

Plants are heterothermic sessile organisms in thermal equilibrium with the environment. Strong temperature variations exceeding lower or higher limits of the thermal optimum for the life of the plant are sensed as thermal stress, cold stress or heat stress (HS), respectively. Thermal stress can compromise the vital functions of the plant more or less severely, depending on the organ affected and its developmental stage [1]. In fact, roots may tolerate lower temperatures than stems and expanded leaves tolerate higher temperatures than the young ones. Thermal stress, in particular HS is one of the main causes of the reduction in crops productivity because the light energy required for photosynthesis results in a considerable increase of the temperature in the exposed tissues [2]. In addition, HS is able to influence growth and biodiversity of forests [3]. This is very important considering the global warming in progress in present years with increasing temperatures and decreasing precipitation with a consequent reduction of wetlands and an increase in areas at risk of desertification [3]. In fact, forests, thanks to their ability to fix carbon dioxide, absorb greenhouse gases and filter anthropogenic pollutants, potentially play a crucial role in the moderation of these changes [3]. Thus, the study of the effects of HS is of great interest for plant biologists.

Plants have evolved different responses to HS to minimize damage and ensure the conservation of cell homeostasis. An intense HS causes a "Heat shock response" (Hsr), which involves the rapid activation of "HS genes" due to the specific transcription factors, named "HS factors" (Hsf). The activity of Hsf induces the synthesis of specific "Heat shock proteins" (Hsp), that act as molecular chaperones involved in plant tolerance to a wide range of stresses [2]. HS can lead to protein denaturation and alteration of the membrane fluidity. This effect may result in high production and accumulation of reactive oxygen species (ROS) causing oxidative stress and, hence, cellular necrosis. On the other hand, low ROS concentrations may act as a second messenger for signal transduction pathways regulating a wide range of cellular functions including programmed cell death (PCD) [4,5]. PCD is an ubiquitous genetically controlled process aimed at eliminating cells that are not necessary or harmful for the proper development of the organism. Among the others, the form of PCD that is the object of major studies and is therefore better known, is the apoptosis of animal cells. Various forms of PCD are also observed in plants, where they are induced by various biotic and abiotic stimuli, including HS [6]. For a long time, selenium (Se) was considered toxic until it was recognized as a micronutrient for humans and other animals [7]. Se is present as selenocysteine in the catalytic site of several selenoproteins involved in important metabolic processes, such as thyroid hormone metabolism, mechanisms of protection from oxidative stress and immune response [8]. In several countries the very low soil concentration of Se causes deficiency in the diet of more than a billion people worldwide [9]. This implies important health problems [10]. Cultivated plants are an important source of Se for humans and livestock. Being chemically analogous to sulphur, Se is absorbed by all plants by sulphate transporters and is sequestered in the form of selenite and selenate [11]. The levels of Se accumulation depend on the abundance of Se in the soil and the levels of the sulphur compounds that compete for absorption [12]. Several attempts were made in order to increase Se content in plants. Changes in the enzymes associated with sulphur metabolism have been widely used to vary Se levels in plants [13]. Recent researches use plant-microbome interactions to increase biofortification with Se and cultivate accumulating plants on seleniferous soils, thus ameliorating soil characteristics for further cultivation. In addition, the biomass of these accumulating plants could be used to enrich the diet of people and their livestock. Finally, given that different species of plants seem able to affect the accumulation of Se from nearby plants and perhaps even their speciation, different co-cultivation techniques could be tested to optimize biofortification with Se of the cultivated plants and their nutritional quality [14]. Despite these studies, until now, there has been no clear evidence of Se essentiality for plants growth. The metal seems to play a dual role: At high doses, it acts as a pro-oxidant agent, causing serious damage to the plant, while low doses can counteract abiotic stress induced by high temperatures, drought, intense light, UV rays, excess of water, salinity and heavy metals [15]. The accumulation of ROS in response to excess of Se may depend on an insufficient presence of antioxidant compounds such as reduced glutathione, thiols, reduced ferredoxin and/or NADPH [16]. These compounds are also involved in the assimilation of Se, thus their concentration can be insufficient to satisfy necessity for Se uptake and at the same time to contrast accumulation of ROS [16]. In contrast, low levels of Se can decrease accumulation of ROS, especially $O_2^{·-}$ and/or H_2O_2, in plants subject to different stresses. Reduction of $O_2^{·-}$ levels can depend on: spontaneous dismutation of $O_2^{·-}$ to H_2O_2 (not catalyzed by the enzyme superoxide dismutase SOD) [17], direct elimination of $O_2^{·-}$ by Se compounds [18], regulation of antioxidant enzymes [15]. However, the mechanisms associated with the protective effect of Se against stresses appears complex and not yet fully understood. In addition to involvement in the mechanism of ROS regulation, a role for Se has been proposed in the inhibition of absorption and translocation of heavy metals. Furthermore, it seems to play a fundamental role in the reconstitution of the cell structures and chloroplasts, and in the recovery of the photosynthetic apparatus after stress [15]. However, an excess of Se could exacerbate the damage to the photosynthetic apparatus and could result in overproduction of starch [19].

Cultured cells are a good experimental material to investigate the responses elicited by HS due to their greater homogeneity compared to complex tissues. Furthermore, this system can be more controlled thus increasing the reproducibility of stress conditions. Previous results showed that in tobacco (*Nicotiana tabacum* cv. Bright-Yellow 2) cultures HS (5 min at 50 °C) induced cell death with apoptotic features, accompanied by oxidative stress and changes in the levels of stress-related proteins [4,6]. In this work we investigated the effect of Se on the responses induced by HS. The obtained results show that Se markedly reduces the effects of HS on cell vitality, cytoplasmic shrinkage, superoxide anion production, membrane lipids peroxidation, activity of caspase-3-like proteases, and the levels of some stress-related proteins (Hsp90, BiP, 14-3-3s, cytochrome c).

2. Results

2.1. HS and Se Effects on Cell Viability and Cytoplasmic Shrinkage

To our knowledge, the effect of Se on plant cultured cells has never been investigated. Thus, to identify the most appropriate Se concentration to use in subsequent experiments, in preliminary experiments we evaluated the effects of different Se concentrations on the accumulation of dead cells induced by HS. Figure 1 shows that in cell cultures not subjected to HS, the percentage of dead cells is very low and does not vary during the experiment. HS determines a progressive accumulation of dead cells, already evident after 3 h of treatment. The results show that there is a progressive protective effect on the appearance of dead cells induced by HS by increasing the concentration of Se up to 1 mM. A further increase in Se concentration does not ameliorate the protective effect but rather seems to reduce it. Therefore, the concentration of 1 mM Se was used in subsequent experiments.

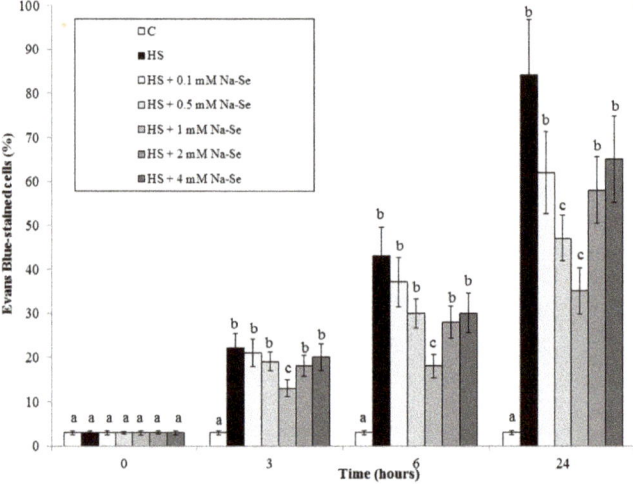

Figure 1. Effect of different Se concentrations on HS-elicited accumulation of dead cells in the cultures. Means ± SD (n ≥ 9) are shown. Different letters show significant differences among treatments at each time (Tukey HSD test, $p \leq 0.05$).

To better characterize the process of death induced by HS, we considered the appearance of cells with shrinked cytoplasm. This morphological modification is presumably caused by destructuration of the cytoskeleton, and in cultured cells is considered an index of PCD with apoptotic features [4,20].

Figure 2 shows that, similarly to the percentage of dead cells, the percentage of control cells with shrinked cytoplasm is very low, constant over time and not influenced by the presence of Se. The treatment with HS leads to a considerable increase in the percentage of cells with shrinked cytoplasm that is in part prevented by Se. At all experimental times it is observed that the percentage of dead cells is slightly higher than that of cells with shrinked cytoplasm, suggesting the presence of different forms of cell death [21].

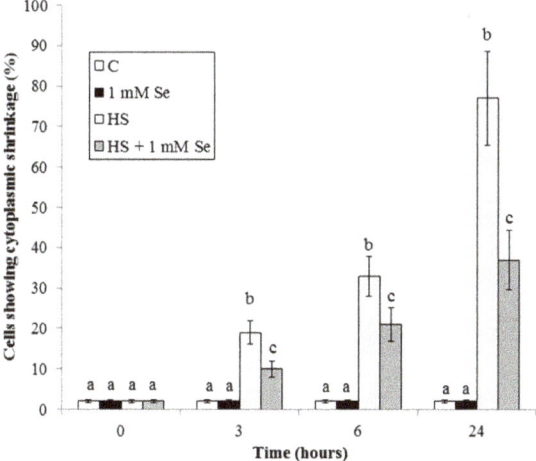

Figure 2. Effect of Se on HS-elicited cytoplasmic shrinkage. Means ± SD (n ≥ 9) are shown. Different letters show significant differences among treatments at each time (Tukey HSD test, $p \leq 0.05$).

2.2. HS and Se Effects on Accumulation of $O_2{\cdot}^-$ and MDA and on Caspase-3-like Activity

$O_2{\cdot}^-$ is a highly reactive ROS, responsible for important oxidative damage [21]. Treatment with HS (Figure 3) causes a progressive accumulation of $O_2{\cdot}^-$. This accumulation is almost totally prevented by Se, at least for the first experimental times (up to 4 h).

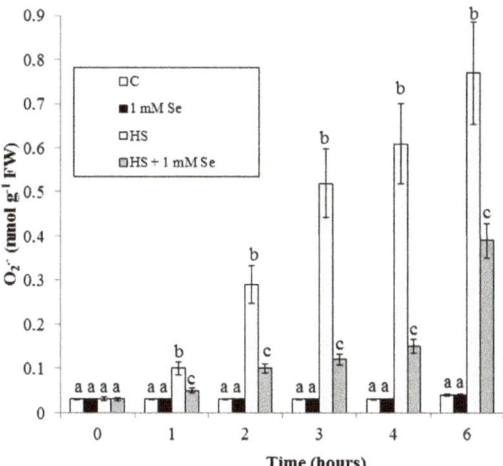

Figure 3. HS and Se effects on O^{2-} accumulation in the culture medium. Means ± SD (n ≥ 9) are shown. Different letters show significant differences among treatments at each time (Tukey HSD test, $p \leq 0.05$).

At cell level, peroxidation of membrane lipids is one of the main damages induced by oxidative stress and its degree was assessed by determining the level of malondialdehyde (MDA), a byproduct of polyunsaturated fatty acids oxidation, which typically originates after oxidative stress [21]. Figure 4 shows that the level of MDA of the control cells is low, constant, and not influenced by Se. At each time, HS causes a considerable production of MDA, significantly decreased by Se.

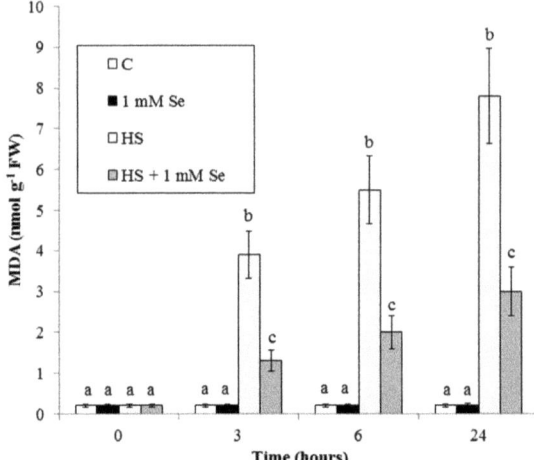

Figure 4. HS and Se effects on MDA accumulation. Means ± SD (n ≥ 9) are shown. Different letters show significant differences among treatments at each time (Tukey HSD test, $p \leq 0.05$).

To characterize further the effect of Se on HS-elicited cell death we analysed caspase3-like proteases activity, another typical PCD marker that often increases during plant PCD. Figure 5 shows that Se strongly reduces the HS-elicited marked increase in this activity.

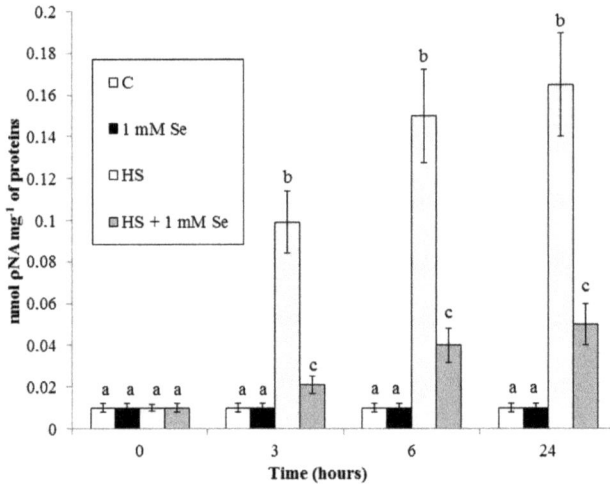

Figure 5. HS and Se effects on caspase-3-like activity. Means ± SD (n ≥ 6) are shown. Different letters show significant differences among treatments at each time (Tukey HSD test, $p \leq 0.05$).

2.3. HS and Se Effects on Stress-related Proteins

Finally, we analysed the effect of HS and Se on the levels of some stress-related proteins by gel blotting. Mitochondrial Hsp 90 are molecular chaperones that control the activity of different substrates. BiP, an Hsp70 present in the endoplasmic reticulum, accumulates under different stress conditions. The regulatory proteins 14-3-3s control many processes of plant cells, including cell death and cytochrome *c* release from the mitochondrion to the cytosol, a marker of apoptotic death in animals and plants [21].

Figure 6 confirms the previously reported effects of HS on the examined proteins [21], and shows that at both times Se diminishes the accumulation of microsomal BiP and almost completely prevents the accumulation of cytosolic 14-3-3s, the reduction of mitochondrial Hsp90 and the release of cytochrome *c* from mitochondria elicited by HS.

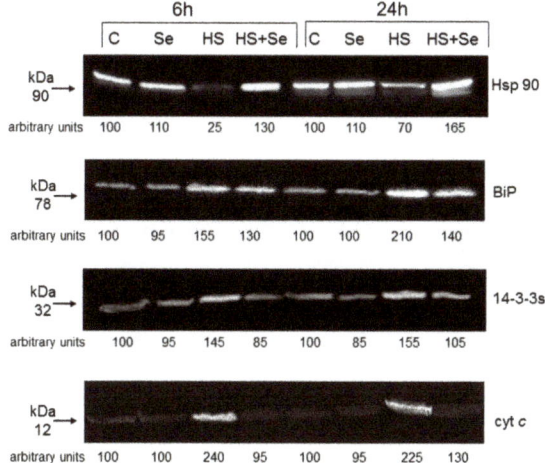

Figure 6. HS and Se effects on the levels of stress-related proteins. (C) control; (Se) cells + 1 mM Se; (HS) HS-treated cells; (HS + Se) HS-treated cells + 1 mM Se. Results of a typical experiment (n = 3) run in duplicate are presented. 50 mg of proteins were run in each lane. An arbitrary value of 100 was assigned to the quantity of immunodecorated protein of the controls.

3. Discussion

The influence of selenium on the HS-elicited responses of tobacco cells was tested by measuring in the absence and presence of Na-selenate the following parameters: cell viability, cytoplasmic shrinkage, superoxide anion production, membrane lipids peroxidation, activity of caspase-3-like proteases, and the levels of some stress-related proteins (Hsp90, BiP, 14-3-3s, cytochrome *c*).

3.1. HS and Se Effects on Cell Viability and Cytoplasmic Shrinkage.

This work (Figures 1 and 2) shows that Se strongly reduces the previously reported eliciting effect of HS on cell death and cytoplasmic shrinkage [6].

At proper concentrations, Se promotes growth, delays plant senescence, and precocious fruit ripening induced by different abiotic stresses, HS included [13]. Leaf and fruit senescence are processes involving programmed cell death, and when not precisely regulated can lead to important decreases in productivity in several horticultural species. The protective effect of Se against inappropriate senescence could be due to its reported ability to reduce respiratory intensity and ethylene production in different plant species [13]. Interestingly, the inhibitor of ethylene production Co^{2+} prevents cell death and cytoplasmic shrinkage induced by HS in tobacco cell cultures [6].

3.2. HS and Se Effects on $O_2^{-\cdot}$ and Malondialdehyde Accumulations.

The HS-elicited oxidative stress with accumulation of $O_2^{\cdot-}$ and malondialdehyde is largely inhibited by Se (Figures 3 and 4). Similar results have been recently obtained in cucumber plants under HS and in *Zea mays* exposed to water stress where stress-induced accumulations of $O_2^{\cdot-}$ and MDA are prevented by Se [22,23]. These results are not surprising. In fact, Se acts as an antioxidant in different plant species under biotic and abiotic stresses. This protective effect of Se depends on the induced higher activity of several antioxidant enzymes and on the increased content of some antioxidant compounds (glutathione and flavonoids) [11,13].

3.3. HS and Se Effects on Caspase-3-like Activity and on Cytochrome c Release

Specific cysteine proteases named caspases are required for the progression of animal apoptosis. In plant cells too there are proteins with similar activity called caspases-like or metacaspases [6]. Se (Figure 5) largely prevents the HS-elicited increase in the activity of these enzymes, reported in our previous work [21]. Another ubiquitous marker of apoptotic-like PCD related to caspases is the release of cytochrome *c* from the mitochondrion [6]. We previously reported induction of cytochrome *c* release by HS [6,21]. Here we show that this release is markedly reduced by Se (Figure 6). These protective effects of Se can be due to its antioxidant effect (ROS are potent regulators of PCD) or to the effect of Se on expression of genes implied in antioxidant activity and defense responses ([11] and see below for further discussion).

3.4. HS and Se Effects on the Levels of Hsp90, BiP and 14-3-3s.

As widely reported, plants evolved a set of responses to deal with HS, that includes changes of biochemical and physiological processes due to modifications of gene expression. These modifications can result in acclimation or adaptation to stress [24]. In this investigation, we studied the HS and Se effects on some stress-related proteins. Our data confirm the previously reported effects of HS on the examined proteins [21] and show that Se diminishes the accumulation of BiP and almost completely prevents the accumulation of 14-3-3s and the reduction of Hsp90 elicited by HS (Figure 6).

To our knowledge, a genomic approach has scarcely been used to study the protective role of Se against stresses. Sun and co-workers performed a comparative proteomics analysis on cucumber plants treated with Cd [25]. Comparing 2-DE gels, these researchers observed several protein spots changed by Se+Cd compared to Cd alone. By MALDI–TOF–MS mass spectrometry, they identified proteins whose relative abundance was significantly reduced by Cd and restored by Se. Among the others, ascorbate oxidase, glutathione-S-transferase and Hsp STI-like expression were strongly reduced by Cd and reincreased by Se [25]. More studies were conducted on the Se effect on the proteome of different plant species. For example, Wang and co-workers by 2-DE gels and MALDI-TOF/TOF mass spectrometry performed a comparative proteomics analysis on the effect of different Se concentrations on rice seedlings [19]. Their results showed that low (non-toxic) Se concentrations up-regulate proteins involved in ROS detoxification and resistance to pathogens such as beta-1,3-glucanase and chitinase. The expression of the same proteins was down-regulated by high (toxic) Se concentrations [19]. Finally, the Se hyperaccumulator *Stanleya pinnata* shows higher expression of genes involved in sulphur uptake and assimilation, antioxidant activities and defense compared to the related secondary Se accumulator *Stanleya albescens* [26]. In particular, the hyperaccumulator species shows a higher expression than the related species for genes encoding Hsp and luminal chaperones such as BiP both in the absence and in the presence of Se [26].

4. Material and Methods

4.1. Cell Culture Growth and Experimental Conditions

Growth of tobacco BY-2 (*Nicotiana tabacum* L. cv Bright-Yellow 2) cells and heat treatment (5 min at 50° C) were performed as described [6]. Na-selenate (Na_2SeO_4) was supplied 10 min before HS.

4.2. Cell Death and Cytoplasmic Shrinkage Assays

Cell death was estimated spectrophotometrically with the vital dye Evans Blue as in Reference [21]. The percentage of cells showing cytoplasmic shrinkage was determined as in Reference [6].

4.3. $O_2^{\cdot-}$ Assay

The $O_2^{\cdot-}$ anion generation was evaluated spectrophotometrically as reduction of XTT to XTT formazan as described [21].

4.4. Proteases Activity and Membrane Lipid Peroxidation

Caspase3-like proteases activity was measured spectrophotometrically with a caspase-3 colorimetric activity assay kit following the manufacturer's instructions (BioVision Research Products, Mountain View, CA 94043, USA) as described [21].

The level of membrane lipid peroxidation was evaluated spectrophotometrically by measuring the content of malondialdehyde, a secondary end product of the oxidation of polyunsaturated fatty acids [21].

4.5. SDS-PAGE and Protein Gel Blots

Cells were collected by gentle centrifugation, frozen in liquid nitrogen and homogenized for 5 min at maximum speed with a Ultra-Turrax T25 device. The cell homogenate was differentially centrifuged to obtain the different fractions (i.e., mitochondrial, microsomal and soluble) for SDS-PAGE analysis as described [21].

Equal amounts of proteins were separated by discontinuous SDS-PAGE (4% stacking, 10% resolving gel) as described [21]. Immunodecorations of cytochrome c, 14-3-3 proteins, Hsp 90 and BiP were perfomed as described [21].

4.6. Statistical Analyses

GraphPad Prism 4 program from GraphPad Software, Inc., San Diego, CA, USA was used to statistically analyse the results. Tukey HSD test, $p \leq 0.05$, was used in the study.

5. Conclusions

To summarize, Se can effectively reduce the effect of HS on cell death with apoptotic features, oxidative stress and levels of stress-related proteins. This protective effect of Se can be due to its direct antioxidant effect and/or to an effect on the expression of genes implied in antioxidant activity and defense responses. In the future, molecular and genomic studies could be valuable to elucidate the mechanisms associated with the protective effect of Se against heat stress in cell cultures as well as in plants.

Author Contributions: Conceptualization, M.M. and R.C.; Funding acquisition, R.C.; Investigation, M.M.; Writing-original draft, M.M.; Writing-review & editing, R.C.

Funding: This research received no external funding

Acknowledgments: Financial support of the University of Milano-Bicocca, Fondo d'Ateneo per la Ricerca, is acknowledged.

Conflicts of Interest: The authors declare that they have no conflict of interest.

References

1. Nejat, N.; Mantri, N. Plant immune system: Crosstalk between responses to biotic and abiotic stresses the missing link in understanding plant defence. *Curr. Issues Mol. Biol.* **2017**, *23*, 1–16. [CrossRef] [PubMed]
2. Kotak, S.; Larkindale, J.; Lee, U.; von Koskull-Döring, P.; Vierling, E.; Scharf, K.D. Complexity of the heat stress response in plants. *Curr. Opin. Plant Biol.* **2007**, *10*, 310–316. [CrossRef] [PubMed]

3. Gielen, B.; Naudts, K.; D'Haese, D.; Lemmens, C.M.H.M.; De Boeck, H.J.; Biebaut, E.; Serneels, L.; Valcke, R.; Nijs, I.; Ceulemans, R. Effects of climate warming and species richness on photochemistry of grasslands. *Physiol. Plant* **2007**, *131*, 251–262. [CrossRef] [PubMed]
4. Vacca, R.A.; de Pinto, M.C.; Valenti, D.; Passarella, S.; Marra, E.; De Gara, L. Production of reactive oxygen species, alteration of cytosolic ascorbate peroxidase, and impairment of mitochondrial metabolism are early events in heat shock-induced programmed cell death in tobacco Bright-Yellow 2 cells. *Plant Physiol.* **2004**, *134*, 1100–1112. [CrossRef] [PubMed]
5. Suzuki, N.; Katano, K. Coordination between ROS regulatory systems and other pathways under heat stress and pathogen attack. *Front. Plant Sci.* **2018**, *9*, 490. [CrossRef] [PubMed]
6. Malerba, M.; Crosti, P.; Cerana, R. Effect of heat stress on actin cytoskeleton and endoplasmic reticulum of tobacco BY-2 cultured cells and its inhibition by Co^{2+}. *Protoplasma* **2010**, *239*, 23–30. [CrossRef] [PubMed]
7. Schwartz, K.; Foltz, C.M. Selenium as an integral part of factor 3 against dietary necrotic liver degeneration. *J. Am. Chem. Soc.* **1957**, *70*, 3292–3293. [CrossRef]
8. Tapiero, H.; Townsend, D.M.; Tew, K.D. The antioxidant role of selenium and seleno-compounds. *Biomed. Pharmacother.* **2003**, *57*, 134–144. [CrossRef]
9. Wu, Z.; Bañuelos, G.S.; Lin, Z.Q.; Liu, Y.; Yuan, L.; Yin, X.; Li, M. Biofortification and phytoremediation of selenium in China. *Front. Plant Sci.* **2015**, *6*, 136. [CrossRef] [PubMed]
10. Roman, M.; Jitaru, P.; Barbante, C. Selenium biochemistry and its role for human health. *Metallomics* **2014**, *6*, 25–54. [CrossRef] [PubMed]
11. Gupta, M.; Gupta, S. An overview of selenium uptake, metabolism, and toxicity in plants. *Front. Plant Sci.* **2017**, *7*, 2074. [CrossRef] [PubMed]
12. Galeas, M.L.; Zhang, L.H.; Freeman, J.L.; Wegner, M.; Pilon-Smits, E.A.H. Seasonal fluctuations of selenium and sulfur accumulation in selenium hyperaccumulators and related non-accumulators. *New Phytol.* **2007**, *173*, 517–525. [CrossRef] [PubMed]
13. Puccinelli, M.; Malorgio, F.; Pezzarossa, B. Selenium enrichment of horticultural crops. *Molecules* **2017**, *22*, 933. [CrossRef] [PubMed]
14. Schiavon, M.; Pilon-Smits, E.A.H. Selenium biofortification and phytoremediation phytotechnologies: A review. *J. Environ. Qual.* **2017**, *46*, 10–19. [CrossRef] [PubMed]
15. Feng, R.; Wei, C.; Tu, S. The roles of selenium in protecting plants against abiotic stresses. *Environ. Exp. Bot.* **2013**, *87*, 58–68. [CrossRef]
16. Feng, R.W.; Wei, C.Y. Antioxidative mechanisms on selenium accumulation in *Pteris vittata* L., a potential phytoremediation plant. *Plant Soil Environ.* **2012**, *58*, 105–110. [CrossRef]
17. Cartes, P.; Jara, A.A.; Pinilla, L.; Rosas, A.; Mora, M.L. Selenium improves the antioxidant ability against aluminium-induced oxidative stress in ryegrass roots. *Ann. Appl. Biol.* **2010**, *156*, 297–307. [CrossRef]
18. Xue, T.L.; Hou, S.F.; Tan, J.A.; Liu, G.L. The antioxidative function of selenium in higher plants: II. Non-enzymatic mechanisms. *Chin. Sci. Bull.* **1993**, *38*, 356–358.
19. Wang, J.D.; Wang, X.; Wong, J.S. Proteomics analysis reveals multiple regulatory mechanisms in response to selenium in rice. *J. Proteom.* **2012**, *75*, 1849–1866. [CrossRef] [PubMed]
20. Hussey, P.J.; Ketelaar, T.; Deeks, M.J. Control of actin cytoskeleton in plant cell growth. *Annu. Rev. Plant Biol.* **2006**, *57*, 109–125. [CrossRef] [PubMed]
21. Malerba, M.; Cerana, R. Role of peroxynitrite in the responses induced by heat stress in tobacco BY-2 cultured cells. *Protoplasma* **2018**, *255*, 1079–1087. [CrossRef] [PubMed]
22. Balal, R.M.; Shahid, M.A.; Javaid, M.M.; Iqbal, Z.; Anjum, M.A.; Garcia-Sanchez, F.; Mattson, N.S. The role of selenium in amelioration of heat-induced oxidative damage in cucumber under high temperature stress. *Acta Physiol. Plant* **2016**, *38*, 158–172. [CrossRef]
23. Bocchini, M.; D'Amato, R.; Ciancaleoni, S.; Fontanella, M.C.; Palmerini, C.A.; Beone, G.M.; Onofri, A.; Negri, V.; Marconi, G.; Albertini, E.; et al. Soil selenium (Se) biofortification changes the physiological, biochemical and epigenetic responses to water stress in *Zea mays* L. by inducing a higher drought tolerance. *Front. Plant Sci.* **2018**, *9*, 389. [CrossRef] [PubMed]
24. Hasanuzzaman, M.; Nahar, K.; Mahabub Alam, M.; Roychowdhury, R.; Fujita, M. Physiological, biochemical, and molecular mechanisms of heat stress tolerance in plants. *Int. J. Mol. Sci.* **2013**, *14*, 9643–9684. [CrossRef] [PubMed]

25. Sun, H.; Dai, H.; Wang, X.; Wang, G. Physiological and proteomic analysis of selenium-mediated tolerance to Cd stress in cucumber (*Cucumis sativus* L.). *Ecotox. Environ. Safe.* **2016**, *133*, 114–126. [CrossRef] [PubMed]
26. Freeman, J.L.; Tamaoki, M.; Stushnoff, C.; Quinn, C.F.; Cappa, J.J.; Devonshire, J.; Fakra, S.C.; Marcus, M.A.; McGrath, S.P.; Hoewyk, D.V.; et al. Molecular mechanisms of selenium tolerance and hyperaccumulation in *Stanleya pinnata*. *Plant Physiol.* **2010**, *153*, 1630–1652. [CrossRef] [PubMed]

© 2018 by the authors. Licensee MDPI, Basel, Switzerland. This article is an open access article distributed under the terms and conditions of the Creative Commons Attribution (CC BY) license (http://creativecommons.org/licenses/by/4.0/).

Article

Morphological and Transcriptome Analysis of Wheat Seedlings Response to Low Nitrogen Stress

Jun Wang [1,2,†], Ke Song [2,3,4,5,6,†], Lijuan Sun [2,3,4,5,6], Qin Qin [2,3,4,5,6], Yafei Sun [2,3,4,5,6], Jianjun Pan [1,*] and Yong Xue [2,3,4,5,6,*]

1. College of Resources and Environmental Sciences, Nanjing Agricultural University, Nanjing 210095, China; 2016103084@njau.edu.cn
2. Eco-Environmental Protection Research Institute, Shanghai Academy of Agricultural Sciences, Shanghai 201403, China; ke.song@wilkes.edu (K.S.); sunlijuan@saas.sh.cn (L.S.); qinqin@saas.sh.cn (Q.Q.); 2014203039@njau.edu.cn (Y.S.)
3. Shanghai Scientific Observation and Experimental Station for Agricultural Environment and Land Conservation, Shanghai Academy of Agricultural Sciences, Shanghai 201403, China
4. Shanghai Environmental Protection Monitoring Station of Agriculture, Shanghai Academy of Agricultural Sciences, Shanghai 201403, China
5. Shanghai Engineering Research Center of Low-Carbon Agriculture (SERLA), Shanghai Academy of Agricultural Sciences, Shanghai 201403, China
6. Shanghai Key Laboratory of Protected Horticultural Technology, Shanghai Academy of Agricultural Sciences, Shanghai 201403, China
* Correspondence: jpan@njau.edu.cn (J.P.); xueyong@saas.sh.cn (Y.X.); Tel.: +86-025-84395329 (J.P.); +86-021-62202594 (Y.X.)
† These authors contributed equally to this work.

Received: 23 February 2019; Accepted: 2 April 2019; Published: 15 April 2019

Abstract: Nitrogen (N) is one of the essential macronutrients that plays an important role in plant growth and development. Unfortunately, low utilization rate of nitrogen has become one of the main abiotic factors affecting crop growth. Nevertheless, little research has been done on the molecular mechanism of wheat seedlings resisting or adapting to low nitrogen environment. In this paper, the response of wheat seedlings against low nitrogen stress at phenotypic changes and gene expression level were studied. The results showed that plant height, leaf area, shoot and root dry weight, total root length, and number under low nitrogen stress decreased by 26.0, 28.1, 24.3, 38.0, 41.4, and 21.2 percent, respectively compared with plants under normal conditions. 2265 differentially expressed genes (DEGs) were detected in roots and 2083 DEGs were detected in leaves under low nitrogen stress (N-) compared with the control (CK). 1688 genes were up-regulated and 577 genes were down-regulated in roots, whilst 505 genes were up-regulated and 1578 were down-regulated in leaves. Among the most addressed Gene Ontology (GO) categories, oxidation reduction process, oxidoreductase activity, and cell component were mostly represented. In addition, genes involved in the signal transduction, carbon and nitrogen metabolism, antioxidant activity, and environmental adaptation were highlighted. Our study provides new information for further understanding the response of wheat to low nitrogen stress.

Keywords: morphological characteristics; transcriptome sequencing; wheat; low nitrogen stress

1. Introduction

Wheat is one of the most widely grown cereal crops all over the world [1]. However, low utilization rate of nitrogen (N) fertilizer severely limits the yield and quality of wheat [2]. Excessive application of nitrogen fertilizer is one of the main ways to ensure crop yield and quality, yet plants can only use about 30% to 40% of the applied nitrogen fertilizer. No less than 40% of the nitrogen fertilizer applied is lost by leaching into the groundwater, lakes, rivers and atmosphere, giving rise to severe pollution [3]. In order to solve this problem, the technology of 'reducing fertilizer and increasing efficiency' has been popularized in our country. Recent studies have shown that insufficient nutrient supply has a serious impact on plant growth [4–6]. Evidence from Jeuffroy et al. indicated that nitrogen deficiency of winter wheat generally can result in slow growth, fewer tillers, and yellowish leaves [7]. The negative effects of low nitrogen on the formation of wheat root morphology have been discussed, including the decreased root length, root number, root surface area, and root dry weight [2,8]. Furthermore, Rose et al. suggested that the developed root architecture has stronger nitrogen uptake capacity, such as greater root length and root surface area [9]. Understanding the morphological response characteristics of wheat seedlings under low nitrogen stress is of critical importance for agricultural fertilization and selection of resistant varieties.

In recent years, transcriptome profiling using next-generation sequencing technologies has been used to study the transcription of genes and the regulation of transcriptional at the overall level [10]. Transcriptome analysis based on Illumina's RNA-sequencing platform in order to explore gene expression in response to nitrogen nutrition stress in plants has been carried out. Wan et al. revealed that wheat amino acid transporters play a vital role in nitrogen transport, response to abiotic stress, and development based on transcriptome analysis [11]. Dai et al. studied how the accumulation of wheat grain storage protein is regulated during grain development in response to nitrogen supply by using transcriptome profiling [12]. Asparagine is considered to be an ideal nitrogen transport molecule, as it plays a major in nitrogen uptake by plant roots [13]. Previous works have found asparagine synthetase genes (AsnS) in *Arabidopsis* [14] and maize [15]. Curci et al. showed that AsnS genes were down-regulated in durum wheat roots and leaves under nitrogen stress [1]. In addition, when plants grow nitrogen-free condition, the genes involved in nitrogen compound metabolism, carbon metabolism, amino acid metabolism, and photosynthesis were down-regulated in roots and leaves [16,17].

Although great progress has been done on the adaptation mechanisms of plants to abiotic stress, little research is available on wheat seedlings response to low nitrogen stress. In this work, besides studying morphological changes between low nitrogen stress and control samples, we focused on transcriptome analysis of wheat seedlings under low nitrogen stress using high-throughput sequencing. Our results provide a basis for understanding wheat's response to low nitrogen stress and the molecular mechanisms that underlie it.

2. Results

2.1. Low Nitrogen Stress Affects Wheat Seedlings Morphology

Wheat seedlings exposure to N- produced a significant growth inhibition compared to control (CK), as observed on plant height, leaf area, shoot dry weight, root dry weight, total root length, and total root number. N- led to a significant ($p < 0.05$) decrease in plant height, leaf area, shoot dry weight, root dry weight, total root length, and total root number by 26.0, 28.1, 24.3, 38.0, 41.4 and 21.2 percent, respectively, compared to CK (Table 1). These results suggest that wheat seedlings are highly sensitive to low nitrogen environments, and low nitrogen stress seriously affected the growth of wheat.

Table 1. Effects of low nitrogen stress on plant height, leaf area, shoot dry weight, root dry weight, total root length, and total root number of wheat seedlings. Different treatments represent: The control (CK), low nitrogen stress (N-). The data are from the average of 15 seedlings, for each parameter, mean values (±standard error) are presented.

Treatment	Plant Height (cm)	Leaf Area (cm²/Plant)	Shoot Dry Weigh (mg/Plant)	Root Dry Weigh (mg/Plant)	Total Root Length (cm/Plant)	Total Root Number
CK	26.58 ± 0.60 [a]	14.70 ± 0.17 [a]	30.77 ± 0.27 [a]	23.17 ± 0.57 [a]	57.83 ± 1.31 [a]	8.50 ± 0.26 [a]
N-	19.67 ± 0.56 [b]	10.57 ± 0.24 [b]	23.29 ± 0.32 [b]	14.37 ± 0.76 [b]	33.90 ± 1.25 [b]	6.70 ± 0.03 [b]

Note: Different letters (a,b) indicate that there are significant differences at 0.05 level according to Tuckey's test.

2.2. Overview of Transcriptome Sequencing Results

The genome-wide transcriptional response to low nitrogen stress in wheat seedlings was investigated by high-throughput RNA-seq. Seedling samples, CK and N-, were sequenced using an Illumina HiSeq platform. Approximately 40.83 to 45.10 million 150 bp paired-end clean reads were obtained from leaf CK and N- samples, respectively, while root CK and N- samples engendered 40.73 to 43.72 million reads, respectively, after adapter trimming and filtering low-quality reads (Table 2). The average leaf Q20, Q30, and GC (Base G + Base C) contents were 95.39%, 89.06%, and 55.00%, respectively. Similarly, the average root Q20, Q30, and GC contents were 95.12%, 88.67%, and 54.00% respectively, with the clean reads of Q20 occupying over 95% of the total reads. These findings attest to the fine quality of the sequencing results. The respective mapped reads information between CK and N- leaf samples was: 84.43% and 84.03% total mapped; 8.06% and 6.50% multiple mapped; 91.94% and 93.50% uniquely mapped, respectively. Between CK and N- root samples 71.01% and 77.41% were total mapped, 6.76% and 5.64% were multiple mapped, while 93.24% and 94.36% were uniquely mapped (Table 2). The transcriptome data was deemed suitable for subsequent analysis.

Table 2. Summary of RNA-seq data and reads mapping. Different treatment represents: The CK and N- of leaf, the CK and N- of root.

Sample		Raw Reads	Clean Reads	Q20 (%)	Q30 (%)	GC (%)	Total Mapped	Multiple Mapped	Uniquely Mapped
Leaf	CK	45,633,102	45,096,752	95.24	88.76	55.38	38,073,404 (84.43%)	3,068,994 (8.06%)	35,004,410 (91.94%)
	N-	41,236,032	40,827,914	95.53	89.35	54.62	34,306,913 (84.03%)	2,229,665 (6.50%)	32,077,248 (93.50%)
Root	CK	41,250,032	40,726,872	95.24	88.94	53.08	28,921,761 (71.01%)	1,953,811 (6.76%)	26,967,950 (93.24%)
	N-	44,308,466	43,721,986	95.00	88.39	54.92	33,846,419 (77.41%)	1,908,786 (5.64%)	31,937,633 (94.36%)

2.3. Low Nitrogen Stress Affects Genes Expression in Wheat Seedlings

Compared with the control, a total of 2265 DEGs were found in roots under low nitrogen stress, of which 1688 genes were up-regulated and 577 genes were down-regulated (Figure 1, Table S1). In a total of 2083 DEGs in leaf transcripts, 505 genes were up-regulated and 1578 genes were down-regulated (Figure 1, Table S2). We observed that a larger number of genes were up-regulated in roots, and a larger number of genes were down-regulated in leaves. Overall, the total number of DEGs in roots were higher than in leaves.

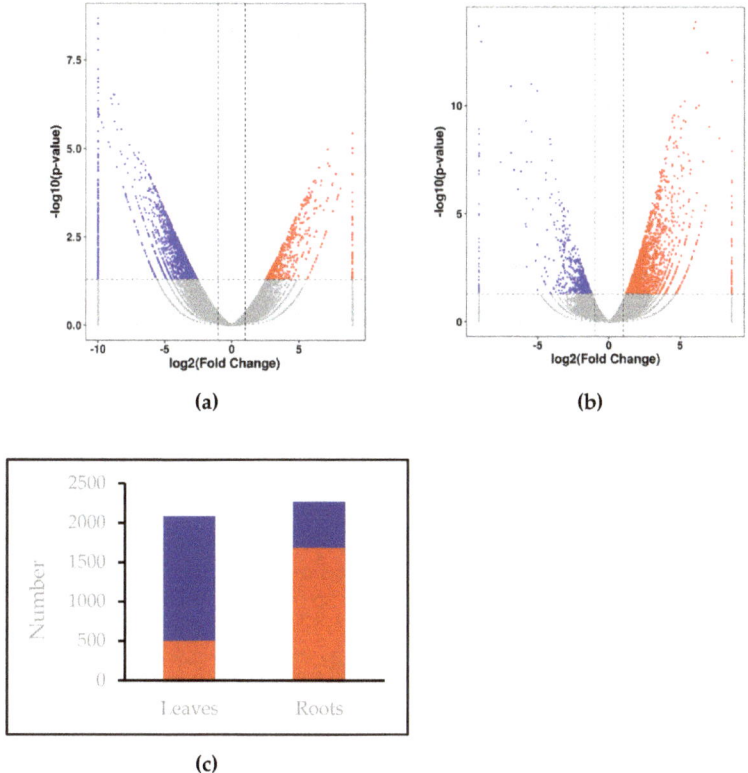

Figure 1. Volcano Plot of differentially expressed genes (DEGs) between the control and low nitrogen stress of wheat seedlings leaves (**a**) and roots (**b**). The two vertical dotted lines are twice of the difference threshold, and the horizontal dotted line represents p-value is 0.05. The red dots indicate the up-regulated genes in this group, the blue dots indicate the down-regulated genes in this group, and the gray dots indicate the non-significant differentially expressed genes. (**c**) Number of genes up-and down-regulated in leaves and roots. The red column indicate the number of up-regulated genes, while the blue column indicate the number of down-regulated genes.

2.4. GO Enrichment Analysis of DEGs

Gene ontology (GO) enrichment analysis was carried out to further characterize the main biological functions of DEGs in wheat seedlings under low nitrogen stress. All the DEGs can be divided into three categories, including biological process, molecular function, and cellular component. Furthermore, the three categories could further be divided into 45 subcategories in the leaf, of which 28 subcategories were significantly ($p < 0.05$) enriched (Figure 2; Table S3). There were ten, ten, and six enriched subcategories belonging to the categories of biological process, molecular function and cellular component, respectively. In biological process, the 'oxidation reduction process' was the most enriched subcategory. 'Oxidoreductase activity' was the most enriched subcategory in the molecular function. Among the cellular components, the three most enriched subcategories were 'cell', 'cell part', and 'organelle'. Other significantly enriched subcategories are shown in Table S3.

Moreover, these three categories could be further divided into 38 subcategories in the root, of which 20 subcategories were significantly ($p < 0.05$) enriched (Figure 3; Table S4). There were ten and nine enriched subcategories belonging to the categories of biological process and molecular function, respectively. Only the 'extracellular region' was significantly enriched in the cellular component.

'Organic acid catabolic process' and 'carbon-nitrogen lyase activity' were the most highly enriched in each of the biological process and molecular function categories, respectively. Other significantly enriched subcategories are shown in Table S4.

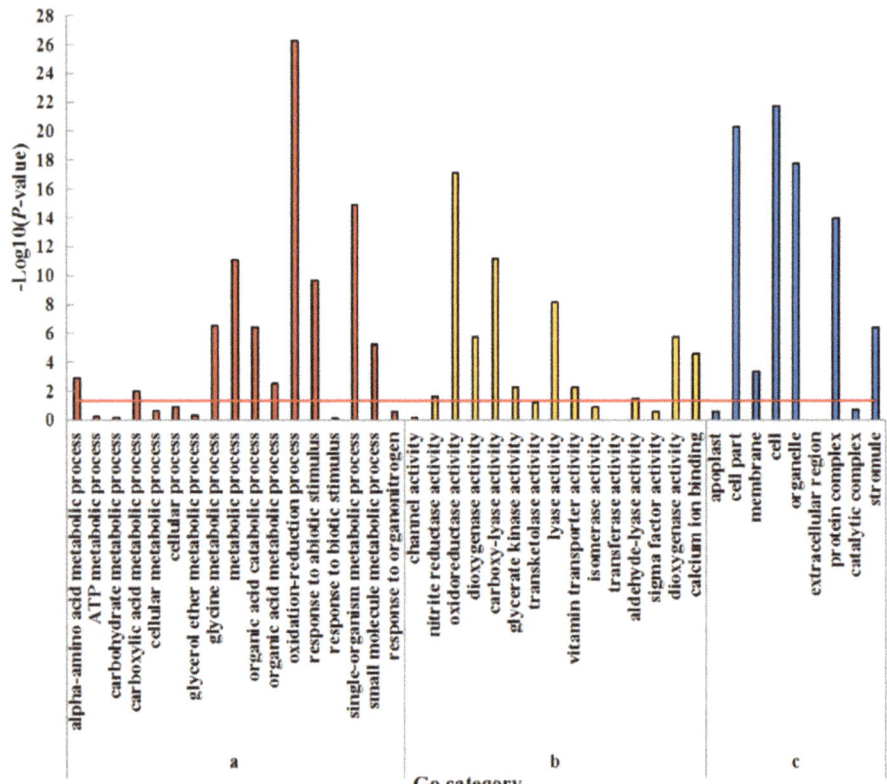

Figure 2. GO enrichment analysis of DEGs in leaves of wheat seedlings. (**a**) Biological process; (**b**) Molecular function; (**c**) Cellular component. The hypergeometric distribution calculates the *p*-value which determines the significance of enrichment, and the red line represents the *p*-value of 0.05.

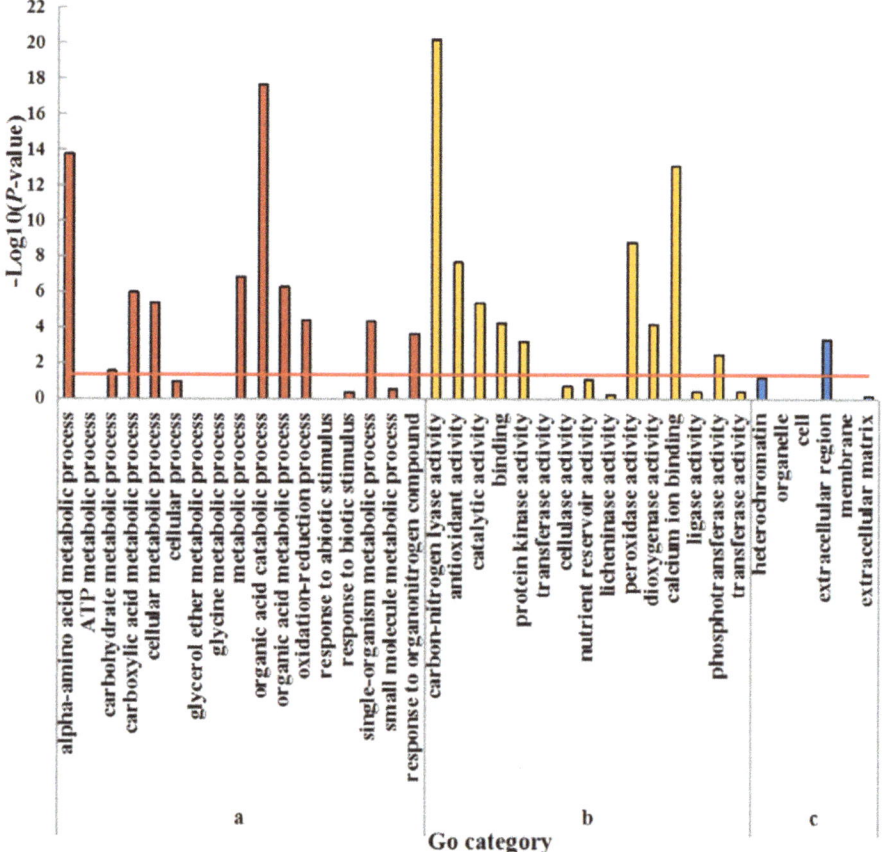

Figure 3. GO enrichment analysis of DEGs in roots of wheat seedlings. (**a**) Biological process; (**b**) Molecular function; (**c**) Cellular component. The hypergeometric distribution calculates the *p*-value which determines the significance of enrichment, and the red line represents the *p*-value of 0.05.

2.5. Kyoto Encyclopedia of Genes and Genomes (KEGG) Enrichment Analysis of DEGs

In order to assign DEGs to cellular pathways, pathway enrichment analysis based on KEGG was performed. The DEGs significantly enriched pathways were calculated using hypergeometric distribution based on the whole genome. There are five KEGG pathway categories: Cellular processes, environmental information processing, genetic information processing, metabolism, and organismal systems. 'Signal transduction' was the only item enriched in environmental information processing in the leaf. In regard to metabolism, 'carbohydrate metabolism' was the most highly overrepresented, followed by 'metabolism of terpenoids and polyketides', 'amino acid metabolism', 'lipid metabolism', 'energy metabolism', 'metabolism of cofactors and vitamins', 'biosynthesis of other secondary metabolites', and 'metabolism of other amino acids'. 'Environmental adaptation' and 'nervous system' were the only items enriched in organismal systems. No enrichment was found in cellular processes and genetic information processing (Figure 4).

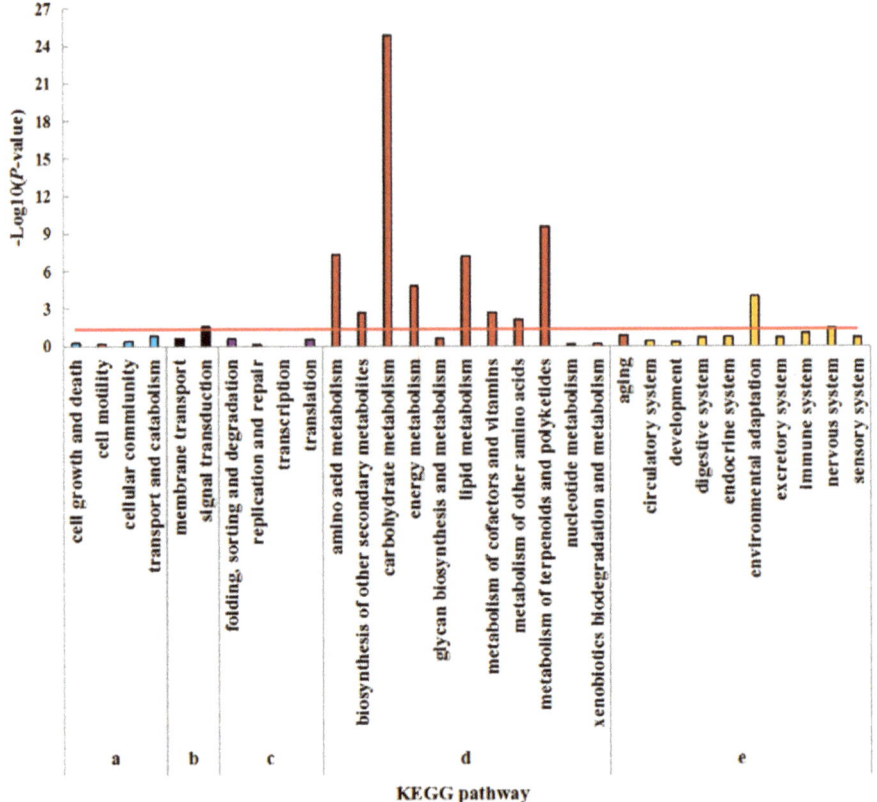

Figure 4. KEGG pathway enrichment analysis of DEGs in leaves under N- compared with CK. (**a**) Cellular processes; (**b**) Environmental information processing; (**c**) Genetic information processing; (**d**) Metabolism; (**e**) Organismal systems. The hypergeometric distribution calculates the *p*-value which determines the significance of enrichment, and the red line represents the *p*-value of 0.05.

Similar KEGG pathway enrichments were found in roots. 'signal transduction' was the only enriched process in environmental information processing. In the category of metabolism, most of the pathways were highly overrepresented, including 'biosynthesis of other secondary metabolites', 'amino acid metabolism', 'lipid metabolism', 'energy metabolism', 'xenobiotics biodegradation and metabolism', 'metabolism of other amino acids', 'carbohydrate metabolism' and 'metabolism of terpenoids and polyketides'. In the category of organismal systems, 'environmental adaptation' was the mostly overrepresented, followed by 'nervous system', 'endocrine system', 'sensory system', and 'digestive system'. No significant enrichment was found in cellular processes and genetic information processing (Figure 5).

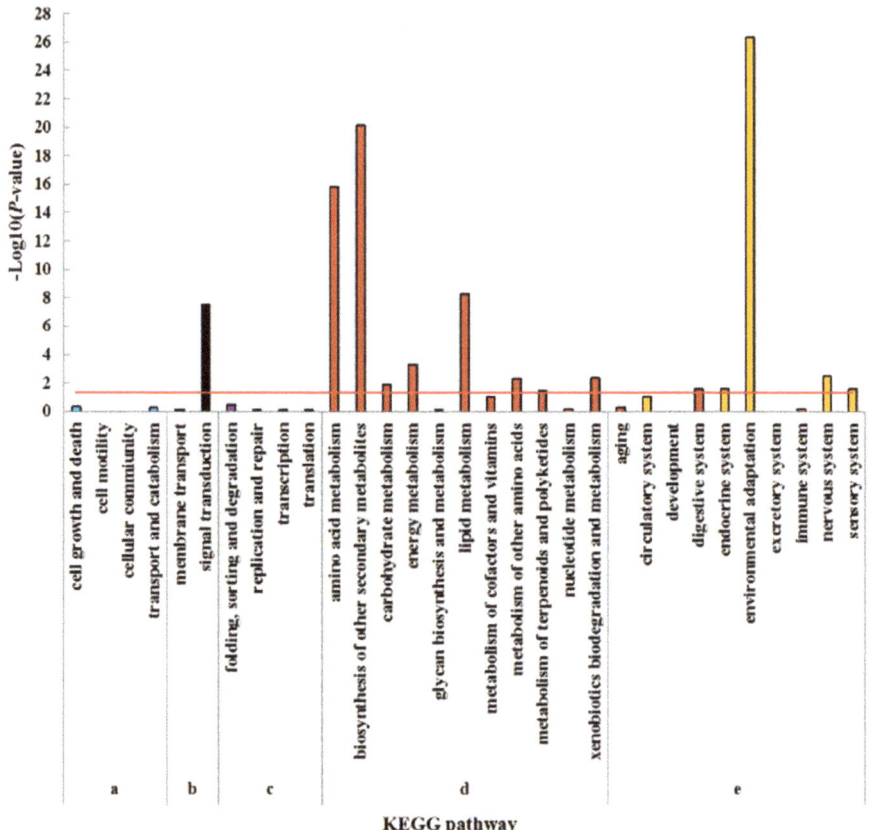

Figure 5. KEGG pathway enrichment analysis of DEGs in roots under N- compared with CK. (**a**) Cellular processes; (**b**) Environmental information processing; (**c**) Genetic information processing; (**d**) Metabolism; (**e**) Organismal systems. The hypergeometric distribution calculates the *p*-value which determines the significance of enrichment, and the red line represents the *p*-value of 0.05.

2.6. Validation of RNA Sequencing Data by Quantitative Real-Time PCR

A total of eight DEGs (four in leaf and four in root) were randomly selected and validated by quantitative real-time PCR (qRT-PCR) to confirm the reliability of our sequencing data. These included genes encoding photosystem II PPD protein 3, MYB transcription factor 78, glycine cleavage system p protein, and catalase in leaves, and WRKY transcription factor, asparagine synthetase, peroxidase, and MYB 33 in roots. The results showed that the average expression levels of four genes in leaf were significantly down-regulated under low nitrogen stress, whereas the average expression levels of four genes in root were significantly up-regulated under low nitrogen stress (Figure 6). The results showed that the expression of these genes measured by qRT-PCR was in good agreement with the RNA-seq results. Consequently, the RNA-seq data we obtained were trustworthy.

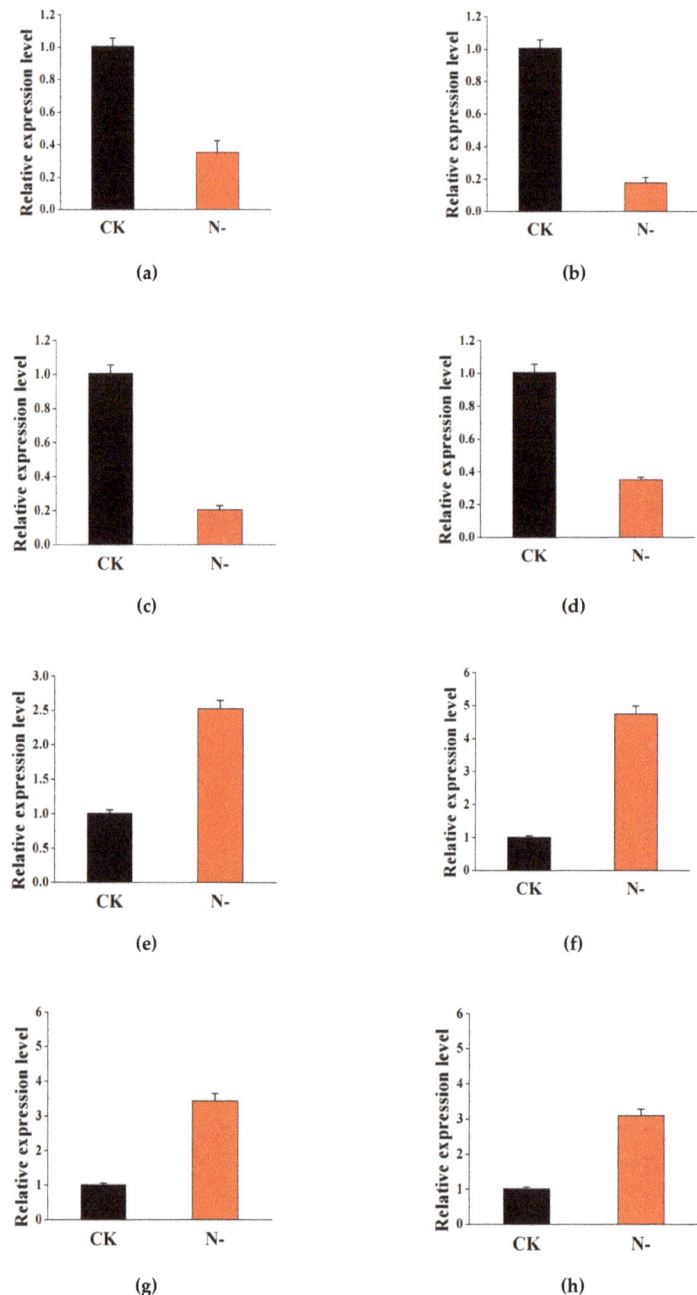

Figure 6. The relative gene expression of eight randomly selected genes examined by qRT-PCR analysis. Different treatments represent: The control (CK), low nitrogen stress (N-). Genes are (**a**) PsbP domain-containing protein 3; (**b**) MYB transcription factor 78; (**c**) Glycine cleavage system p protein; (**d**) Catalase; (**e**) WRKY transcription factor; (**f**) Asparagine synthetase; (**g**) Peroxidase; (**h**) MYB 33. Data represent the mean ± SE ($n = 3$).

3. Discussion

3.1. Response to Low Nitrogen Stress by Morphological Changes

Nitrogen is an essential nutrient for plant growth and development [17]. Insufficient nitrogen supply adversely influences morphology, limits growth, and decreases biomass in wheat [18,19]. Most plants manifest prominent changes in their leaves and roots when grown under low phosphorus or nitrogen conditions. Plants rely on morphological changes to adapt to nutrient stress, a common finding on plants grown under nutrient stress conditions [20–22]. Our results further confirmed that low nitrogen stress inhibited wheat growth. We recorded a remarkable response to low nitrogen stress in wheat seedlings. Low nitrogen treatment significantly decreased wheat seedlings growth parameters, including plant height, leaf area, shoot and root dry weight, and total root length and number. The changes in these parameters suggested that low nitrogen stress has a serious impact on wheat seedlings growth.

3.2. Potential DEGs Play Important Roles in Low Nitrogen Tolerance in Wheat Seedlings

Abiotic stress triggers dramatic molecular responses in plants. In recent years, the molecular mechanism of plant response to abiotic stress has attracted wide attention [23–26]. In order to obtain the molecular mechanism of wheat seedlings response to low nitrogen stress, RNA-seq analysis was performed. 4348 differentially expressed genes were obtained from the whole wheat seedling in all, and among which 2265 genes (1688 up-regulated and 577 down-regulated) were from roots, and 2083 genes (505 up-regulated and 1578 down-regulated) were from leaves. Our results showed that the gene expression of wheat seedlings changed greatly under low nitrogen stress. Changes in gene expression can lead to changes in the corresponding biological processes. GO enrichment analysis is helpful to highlight the main biological processes in response to stress environment. For example, 'nitrogen compound metabolism', 'carbon metabolism', and 'photosynthesis' were mostly enriched in durum wheat under nitrogen starvation [17]. 'Metabolic process', 'cellular process', and 'transport' were enriched in rice roots and shoots under nitrogen-free conditions [16]. Our results detected 'oxidation-reduction process' and 'metabolic process' as highly enriched in wheat leaves subjected to low nitrogen stress. Genes involved in 'carbon–nitrogen lyase activity' and 'organic acid catabolic process' were significantly enriched in wheat seedlings roots. Bi et al. indicated that DEGs associated with these processes might play essential roles in *Arabidopsis thaliana* adaptation to low nitrogen stress [27]. Therefore, the response of these biological process genes enhances the adaptability of wheat to low nitrogen stress.

3.3. Amino Acid Metabolism, Lipid Metabolism, Energy Metabolism, and Signal Transduction Pathway Play Important Roles Under Low Nitrogen Stress

KEGG pathway analysis can help us to further understand the biological functions of genes and how these genes interact [28]. Previous studies have revealed that many genes participate in nitrogen deficiency or nitrogen-free condition resistance via various amino acid metabolism, lipid metabolism, energy metabolism, and signal transduction pathways [29–31]. For example, DEGs associated with amino acid metabolism pathway were mostly represented in sorghum roots under low nitrogen stress, which play a key role in nitrogen uptake and transformation [32]. Protein kinases (PK) are widely involved in the signal transduction and were shown to respond to nitrogen deficiency in *Arabidopsis thaliana* roots and leaves [17]. In this study, genes related to 'amino acid metabolism pathway', 'lipid metabolism pathway', 'energy metabolism pathway', and 'signal transduction pathway' were also identified both in leaves and roots. These results support previous findings that these pathways can form a close-knit signaling network and play vital roles in low nitrogen stress tolerance in plants [28].

3.4. Some Candidate Genes for Plant Low Nitrogen Stress Tolerance Breeding

PsbP protein is an extrinsic component of photosystem II (PSII) and participates in crucial processes such as calcium-ion binding and photosynthesis [33,34]. Studies have shown that genes encoding PsbP protein can serve as signal response factors to various unfavorable external environments [35,36]. In our study, PsbP domain-containing protein 3 (PPD3) gene, a member of the PsbP gene family, was predominantly down-regulated in wheat leaves under low nitrogen stress., Before this, no studies have found that PPD3 gene is regulated in wheat under low nitrogen or nitrogen-free stress. Our results provide a reference for future research.

Transcription factors (TFs) are important factors involved in abiotic stress regulation in plants, and, among them, MYB, WRKY, bHLH, and bZIP families were identified responding to nitrogen stress [37,38]. For example, genes encoding WRKY TFs were up-regulated in rice under nitrogen deficiency [16], MYB TFs were up-regulated in durum wheat under nitrogen starvation [17], and also revealed that NF-Y TFs were induced in wheat under low nitrogen conditions, which play roles in nitrogen uptake and grain yield [39]. Some TFs regulate the pivotal cell process in the response to nitrogen deficiency. Zhang et al. reported that several MYB TFs are associated with the cell development and cell cycle in wheat during nitrogen stress [40]. In this paper, MYB and WRKY were the most prominent. In regard to WRKY family genes, 17 genes were up-regulated in roots under low nitrogen stress, 40 MYB family genes were detected in both the leaves and roots. They were down-regulated in leaves, but up-regulated in roots..Our results suggest that these transcription factors detected in response to low nitrogen stress may contribute to the adaptation or resistance of wheat seedlings to low nitrogen condition.

Asparagine (Asn) play a vital role in nitrogen metabolism, transport, and storage [41]. Asparagine biosynthesis is catalyzed by asparagine synthase (AsnS) in higher plants [42]. Recent studies showed that the AsnS gene family also serves as a response factor to nitrogen and other abiotic stresses [17]. Asparagine synthetase 1 (AS1), a member of the AsnS family, was shown in rice to be expressed mainly in roots in an NH_4^+-dependent manner [43]. In the present work, several genes belonging to the AsnS gene family were mostly up-regulated in wheat roots under low nitrogen stress. The significant differential expression of AsnS genes in wheat roots under low nitrogen condition may contribute to the synthesis of asparagine, and promote the assimilation and distribution of nitrogen.

Reactive oxygen species (ROS) often accumulates in plants under various abiotic stresses, which leads to the oxidative damage of many cell structures and components [44,45]. In order to protect themselves from oxidative damage, plants have developed an antioxidant defense system consisting of a variety of enzymes. Superoxide dismutase (SOD), peroxidases (POD), and catalases (CAT) are common antioxidant enzymes. Lian et al. have reported that antioxidant defense systems can be induced under low nitrogen stress, and improve the adaptability to low nitrogen environment, by increasing the gene expression levels of SOD, POD, and CAT in rice [46]. Bi et al. also found that several detoxification-associated genes were detected in *Arabidopsis* under nitrogen stress [27]. In the present study, plenty of POD genes were up-regulated in wheat roots under low nitrogen stress, and several CAT genes were down-regulated in leaves. These genes might improve the adaptability of wheat seedling to low nitrogen environments.

4. Material and Methods

4.1. Plant Material and Growth Conditions

For this study, wheat cultivar Wanmai No. 52 grown in large areas in China, was selected. The wheat seedlings were hydroponically cultured in a greenhouse. After selecting full and uniform seeds, they were sterilized with ethanol (75% *v/v*) for one min and then washed three times in distilled water, soaked in distilled water and placed in an artificial climate incubator at 25 °C for 24 h without light. Uniformly germinated seeds were selected and placed on a moistened germination paper, and then cultured in a greenhouse at 25 ± 3 °C. At about 5 cm growth height, the robust seedlings were transferred to plastic boxes containing a proper volume of hydroponic nutrient solution (pH 6.0), and were grown afterwards in an incubator at 20/15 °C (day/night) under a 12 h photoperiod until they showed two fully expanded leaves. During the growth period, seedlings were sprayed with distilled water on time. The growth of wheat is shown in Figure S1.

4.2. Experimental Design

The basic nutrient solution was an improved Hoagland nutrient solution. The major elements of nutritional solution are as follows: 4 mM $Ca(NO_3)_2 \cdot 4H_2O$, 5 mM KNO_3, 1 mM NH_4NO_3, 1 mM KH_2PO_4, 2 mM $MgSO_4 \times 7H_2O$, 0.02 mM $FeSO_4 \times 7H_2O$ + EDTA(Na). The micronutrients of nutritional solution are as follows: 0.005 mM KI, 0.1 mM H_3BO_3, 0.15 mM $MnSO_4$, 0.05 mM $ZnSO_4$, 0.001 mM Na_2MoO_4, 0.15×10^{-3} mM $CuSO_4$, 0.19×10^{-3} mM $CoCl_2$. Two-leaf stage seedlings were transferred to different conditions: Standard total nitrogen nutrition ($Ca(NO_3)_2 \cdot 4H_2O$ + KNO_3 + NH_4NO_3 2 mM, labeled CK), and low-nitrogen stress ($Ca(NO_3)_2 \cdot 4H_2O$ + KNO_3 + NH_4NO_3 0.2 mM, labeled N-). Except for the total nitrogen concentration, the other components of the solution under low-nitrogen stress were identical to those of the control. Each treatment was applied to 18 plants and repeated 3 times. The incubator was placed in a greenhouse with a 12 h photoperiod and the nutrient solution was renewed every three days. After ten days of treatments, roots and leaves were sampled from CK and N- wheat seedlings, and immediately frozen in liquid nitrogen, and then refrigerate it in the refrigerator with −80 °C until RNA extraction.

4.3. Determination of Morphological Parameters

Morphological parameters (plant height, leaf area, shoot dry weight, root dry weight, total root length and number) were determined from harvested wheat seedlings. Plant height and leaf area were determined following Wan [47]. Leaf area was calculated according to the following formula:

$$\text{Leaf area (cm}^2\text{)} = \text{leaf length} \times \text{leaf width}/1.2$$

Roots were scanned with a scanner, and images were analyzed using Win Rhizo to determine total root length and total root number according to Boris [48]. Seedlings dry weight was determined according to Qiu [49].

4.4. Isolation of Total RNA, cDNA Library Construction, and Illumina Sequencing

Total RNA was extracted from roots and leaves by the TRIzol Reagent (Invitrogen, Carlsbad, CA, USA) according to the manufacturer's instructions. Agilent 2100 Bioanalyzer (Agilent Technologies, Santa Clara, CA, USA) was used to determine RNA concentration and purity. RNA integrity was examined by running gel electrophoresis in 1% (w/v) agarose gel (Invitrogen, CA, USA). The next step is to enrich mRNA from total RNA using Oligo (dT)-conjugated magnetic beads. mRNA was enriched from total RNA using mRNA fragmentation was conducted by ion interruption, giving rise to mRNA fragments of about 200 bp. The first strand of cDNA was synthesized using reverse transcriptase and random primers using the above fragments as templates, and the second strand was synthesized using DNA polymerase I and RNase H. The end of the fragment was connected with the adapter. Then the products were purified and concentrated by polymerase chain reaction (PCR) to form the final cDNA libraries. The libraries were paired-end (PE)-sequenced using next-generation sequencing based on the Illumina HiSeq platform. Sequencing was conducted by Shanghai Personal Biotechnology Co., Ltd. (Shanghai, China). The raw reads were submitted to the NCBI Sequence Read Archive (accession number: PRJNA528563).

4.5. Data Filtering and Mapping of Illumina Reads

To obtain high-quality reads for subsequent analysis, FastQC quality control tool (http://www.bioinformatics.babraham.ac.uk/projects/fastqc/) was used to evaluate the quality of the quality of

RNA-seq raw reads. The reads were then filtered using Cutadapt. The process included removing adapter sequences and trimming the bases with a quality score less than Q20 using a 5-bp 3' to 5' window. Final reads that are less than 50 bp in length or do not require bases are discarded. A reference genome index was established using Bowtie2 [50], and clean reads were mapped to the reference genome (http://www.ensembl.org/index.html) using Tophat2 [51].

4.6. Enrichment Analysis of DEGs

The DESeq R package was performed to analyze the differential expression in each tissue. At first, we mapped high-quality reads data to the reference genome of wheat (http://www.ensembl.org/) to calculate the number of reads mapped to each gene in each sample. Then these raw read counts were normalized to reads per kilo bases per million reads (RPKM). We use a false discovery rate of 0.05 as the threshold for judging the significance of DEGs. We performed enrichment analysis using GO (http://geneontology.org/) and the KEGG (http://www.genome.jp/kegg). DEGs terms and pathways were calculated using a hypergeometric distribution algorithm with wheat reference genome (Table S5) (http://www.ensembl.org/) as background.

4.7. Quantitative Real-Time PCR Validation

In order to validate the reliability of DEGs obtained from RNA-seq, four DEGs were randomly selected from leaves and roots, respectively, for qRT-PCR analysis. *RLI* (Ta2776) was used as an internal control. Gene-specific primer pairs were designed using Primer Premier 5.0 software (Premier Biosoft International), and primer information is shown in Table S6. Total RNA from each tissue was extracted as described above. Two micrograms RNA was reverse-transcribed into cDNA using the iScriptTM advanced cDNA Synthesis Kit (Promega, Madison, WI, USA) following RNase-free DNase I (Promega, Madison, WI, USA) treatment. Standard curve for each gene was prepared with several dilutions of cDNA. The qRT-PCR was carried out using SYBR® Green PCR Master Mix (Roche, CH) in a Rotor-Gene 3000 Real Time system (Qiagen, Hilden, Germany). Quantitative PCR reactions cycling conditions were performed as follows: 95 °C for 5 min, followed by 40 cycles at 95 °C for 15 s, 60 °C for 30 s. The relative expression value of the different genes was calculated using $2^{-\Delta\Delta Ct}$ method [52]. The experiment was performed using three biological replicates.

4.8. Statistical Analysis

Statistical analysis of plant morphological data was conducted using SPSS 19.0 (IBM, Chicago, IL, USA). Statistical results were obtained by one-way analysis of variance (ANOVA) followed by Tuckey's test to evaluate significant treatment effects at significance of $p < 0.05$. Data presented are means ± standard errors.

5. Conclusions

Our study provides a more comprehensive understanding of the morphological changes and the DEGs in wheat roots and leaves under low nitrogen stress. It was observed that plant height, leaf area, shoot dry weight, root dry weight, total root length and total root number of wheat seedlings were decreased significantly in under nitrogen stress. 2265 and 2083 DEGs were detected in roots and leaves, respectively.In regard to up-regulation and down-regulation, 1688 genes were up-regulated, and 577 genes were down-regulated in roots, while 505 genes were up-regulated, and 1578 genes were down-regulated in leaves. Furthermore, the classification functional enrichment and metabolic pathways of DEGs were shown in this paper. Several key genes and TFs involved in the signal transduction, carbon and nitrogen metabolism, antioxidant activity, and environmental adaptation were identified. Our results will provide useful information to the further research about wheat response to low nitrogen stress.

Supplementary Materials: The following are available online at http://www.mdpi.com/2223-7747/8/4/98/s1; Figure S1: View of wheat seedlings growth in CK and N- groups. Table S1: All expression datasets of DEGs under low N stress in roots. Table S2: All expression datasets of DEGs under low N stress in leaves. Table S3: GO enrichment analysis for DEGs in wheat seedlings under low nitrogen stress (leaf). Table S4: GO enrichment analysis for DEGs in wheat seedlings under low nitrogen stress (root). Table S5: Reference genome information. Table S6: Primer information for qRT-PCR.

Author Contributions: J.W. and K.S. performed the experiments and wrote the manuscript. L.S., Q.Q. and Y.S. helped to revise the manuscript. J.P. and Y.X. involved in designing the research and revised the manuscript. All authors read and approved the manuscript.

Funding: This study was funded by the National Key Research and Development Program of China (2018YFD0200500), Shanghai Science and Technology Commission's Yangtze River Delta Science and Technology Joint Research Project (17295810062), Shanghai Academy of Agricultural Sciences Outstanding Team Plan (2017(A-03)), Shanghai Science and Technology Commission's Key Project in Social Development Field (17DZ1202301), Natural Science Foundation of Shanghai (16ZR1431100, 17ZR1431200), and Discipline Construction Project of Shanghai Academy of Agricultural Sciences (Promotion 2019 (22)).

Conflicts of Interest: The authors declare no conflicts of interest.

References

1. Curci, P.L.; Bergès, H.; Marande, W.; Maccaferri, M.; Tuberosa, R.; Sonnante, G. Asparagine synthetase genes (AsnS1, and AsnS2) in durum wheat: Structural analysis and expression under nitrogen stress. *Euphytica* **2018**, *214*, 36. [CrossRef]
2. Zhang, D.; Zhang, Y.; Yang, W.; Miao, G. Biological response of roots in different spring wheat genotypes to low nitrogen stress. *Acta Agron. Sin.* **2006**, *32*, 1349–1354.
3. Frink, C.R.; Waggoner, P.E.; Ausubel, J.H. Nitrogen fertilizer: Retrospect and prospect. *Proc. Natl. Acad. Sci. USA* **1999**, *96*, 1175–1180. [CrossRef] [PubMed]
4. Hill, J.O.; Simpson, R.J.; Moore, A.D. Morphology and response of roots of pasture species to phosphorus and nitrogen nutrition. *Plant Soil* **2006**, *286*, 7–19. [CrossRef]
5. Gruber, B.D.; Giehl, R.F.; Friedel, S.; Von, W.N. Plasticity of the arabidopsis root system under nutrient deficiencies. *Plant Physiol.* **2013**, *163*, 161–179. [CrossRef] [PubMed]
6. Trubat, R.; Cortina, J.; Vilagrosa, A. Plant morphology and root hydraulics are altered by nutrient deficiency in *Pistacia lentiscus* (L.). *Trees* **2006**, *20*, 334–339. [CrossRef]
7. Jeuffroy, M.H.; Bouchard, C. Intensity and duration of nitrogen deficiency on wheat grain number. *Crop Sci.* **1999**, *39*, 1385–1393. [CrossRef]
8. Xue, Y.F.; Zhang, W.; Liu, D.Y.; Yue, S.C.; Zou, C.Q. Effects of nitrogen management on root morphology and zinc translocation from root to shoot of winter wheat in the field. *Field Crops Res.* **2014**, *161*, 38–45. [CrossRef]
9. Rose, T.J.; Impa, S.M.; Rose, M.T.; Pariasca-Tanaka, J.; Mori, A.; Heuer, S.; JohnsonBeebout, S.E.; Wissuwa, M. Enhancing phosphorus and zinc acquisition efficiency in rice: A critical review of root traits and their potential utility in rice breeding. *Ann. Bot.* **2013**, *112*, 331–345. [CrossRef]
10. Diao, J.; Liu, H.; Hu, F.; Li, L.; Wang, X.; Gai, C. Transcriptome analysis of immune response in fat greenling (*Hexagrammos otakii*) against Vibrio harveyi infection. *Fish Shellfish Immunol.* **2018**, *84*, 937–947. [CrossRef]
11. Wan, Y.; King, R.; Mitchell, R.A.C.; Hassani-Pak, K.; Hawkesford, M.J. Spatiotemporal expression patterns of wheat amino acid transporters reveal their putative roles in nitrogen transport and responses to abiotic stress. *Sci. Rep.* **2017**, *7*, 5461–5474. [CrossRef]
12. Dai, Z.W.; Plessis, A.; Vincent, J.; Duchateau, N.; Besson, A.; Dardevet, M.; Prodhomme, D.; Gibon, Y.; Hilbert, G.; Pailloux, M.; et al. Transcriptional and metabolic alternations rebalance wheat grain storage protein accumulation under variable nitrogen and sulfur supply. *Plant J.* **2015**, *83*, 326–343. [CrossRef]
13. Kirkman, M.A.; Miflin, B.J. The nitrate content and amino acid composition of the xylem fluid of spring wheat throughout the growing season. *J. Sci. Food Agric.* **1979**, *30*, 653–660. [CrossRef]
14. Lam, H.M.; Hsieh, M.-H.; Coruzzi, G. Reciprocal regulation of distinct asparagine synthetase genes by light and metabolites in Arabidopsis thaliana. *Plant J.* **1998**, *16*, 345–353. [CrossRef]
15. Todd, J.; Screen, S.; Crowley, J. Identification and characterization of four distinct asparagine synthetase (*AsnS*) genes in maize (*Zea mays* L.). *Plant Sci.* **2008**, *175*, 799–808. [CrossRef]
16. Yang, S.Y.; Hao, D.L.; Song, Z.Z.; Yang, G.Z.; Wang, L.; Su, Y.H. RNA-Seq analysis of differentially expressed genes in rice under varied nitrogen supplies. *Gene* **2015**, *555*, 305–317. [CrossRef]

17. Curci, P.L.; Cigliano, R.A.; Zuluaga, D.L.; Janni, M.; Sanseverino, W.; Sonnante, G. Transcriptomic response of durum wheat to nitrogen starvation. *Sci. Rep.* **2017**, *7*, 1176–1190. [CrossRef]
18. Forde, B.; Lorenzo, H.; Powlson, D.S. The nutritional control of root development. *Plant Soil* **2001**, *232*, 51–68. [CrossRef]
19. Werf, A.V.D.; Nuenen, M.V.; Visser, A.J.; Lambers, H. Contribution of physiological and morphological plant traits to a species' competitive ability at high and low nitrogen supply. *Oecologia* **1993**, *94*, 434–440. [CrossRef]
20. Rga, B.; Mensink, M. *The Influence of Nitrogen Availability on Growth Parameters of Fast- and Slow-Growing Perennial Grasses*; Blackwell Science: Hoboken, NJ, USA, 1991; pp. 161–168.
21. Boot, R.G.A.; Mensink, M. Size and morphology of root systems of perennial grasses from contrasting habitats as affected by nitrogen supply. *Plant Soil* **1990**, *129*, 291–299. [CrossRef]
22. Schippers, P.; Olff, H. Biomass partitioning, architecture and turnover of six herbaceous species from habitats with different nutrient supply. *Plant Ecol.* **2000**, *149*, 219–231. [CrossRef]
23. Li, X.J.; Guo, C.J.; Lu, W.J.; Duan, W.W.; Zhao, M.; Ma, C.Y. Expression pattern analysis of zinc finger protein genes in wheat (*Triticum aestivum* L.) under phosphorus deprivation. *J. Int. Agric.* **2014**, *13*, 1621–1633. [CrossRef]
24. Li, C.; Li, Y.; Bai, L.; Chaoxing, H.E.; Xianchang, Y.U. Dynamic expression of miRNAs and their targets in the response to drought stress of grafted cucumber seedlings. *Hortic. Plant J.* **2016**, *2*, 41–49. [CrossRef]
25. Jun, H.E.; Ren, Y.; Wang, Y. Root morphological and physiological responses of rice seedlings with different tolerance to cadmium stress. *Acta Ecol. Sin.* **2011**, *31*, 522–528.
26. Li, H.; Hu, T.; Amombo, E.; Fu, J. Transcriptome profilings of two tall fescue (Festuca arundinacea) cultivars in response to lead (Pb) stress. *BMC Genom.* **2017**, *18*, 145. [CrossRef]
27. Bi, Y.M.; Wang, R.L.; Tong, Z.; Rothstein, S.J. Global transcription profiling reveals differential responses to chronic nitrogen stress and putative nitrogen regulatory components in *Arabidopsis*. *BMC Genom.* **2007**, *8*, 281. [CrossRef]
28. Kanehisa, M.; Goto, S.; Kawashima, S.; Okuno, Y.; Hattori, M. The KEGG resource for deciphering the genome. *Nucleic Acids Res.* **2004**, *32*, D277–D280. [CrossRef]
29. Stitt, M. Nitrate regulation of metabolism and growth. *Curr. Opin. Plant Biol.* **1999**, *2*, 178–186. [CrossRef]
30. Lam, H.M.; Coschigano, K.T.; Oliveira, I.C.; Melo-Oliveira, R.; Coruzzi, G.M. The molecular-genetics of nitrogen assimilation into amino acids in higher plants. *Annu. Rev. Plant Physiol. Plant Mol. Biol.* **1996**, *47*, 569–593. [CrossRef]
31. Scheible, W.R.; GonzalezFontes, A.; Lauerer, M.; MullerRober, B.; Caboche, M.; Stitt, M. Nitrate acts as a signal to induce organic acid metabolism and repress starch metabolism in tobacco. *Plant Cell* **1997**, *9*, 783–798. [CrossRef]
32. Gelli, M.; Duo, Y.; Konda, A.; Zhang, C.; Holding, D.; Dweikat, I. Identification of differentially expressed genes between sorghum genotypes with contrasting nitrogen stress tolerance by genome-wide transcriptional profiling. *BMC Genom.* **2014**, *15*, 179. [CrossRef] [PubMed]
33. Hall, M.; Kieselbach, T.; Sauer, U.H.; Schröder, W.P. Purification, crystallization and preliminary X-ray analysis of PPD6, a PsbP-domain protein from Arabidopsis thaliana. *Acta Crystallogr.* **2012**, *68*, 278–280.
34. Bricker, T.M.; Roose, J.L.; Zhang, P.; Frankel, L.K. The PsbP family of proteins. *Photosynth. Res.* **2013**, *116*, 235–250. [CrossRef] [PubMed]
35. Liu, J.; Yang, H.; Lu, Q.; Wen, X.; Chen, F.; Peng, L. PsbP-domain protein1, a nuclear-encoded thylakoid lumenal protein, is essential for photosystem I assembly in Arabidopsis. *Plant Cell* **2012**, *24*, 4992–5006. [CrossRef] [PubMed]
36. Fey, V.; Wagner, R.; Braütigam, K.; Wirtz, M.; Hell, R.; Dietzmann, A. Retrograde plastid redox signals in the expression of nuclear genes for chloroplast proteins of *Arabidopsis thaliana*. *J. Biol. Chem.* **2005**, *280*, 5318–5328. [CrossRef] [PubMed]
37. Singh, K.B. Transcription factors in plant defense and stress responses. *Curr. Opin. Plant Biol.* **2002**, *5*, 430–436. [CrossRef]
38. Wu, W.S.; Chen, B.S. Identifying stress transcription factors using gene expression and TF-gene association data. *Bioinform. Biol. Insights* **2007**, *1*, 137–145. [CrossRef]
39. Qu, B.; He, X.; Wang, J.; Zhao, Y.; Teng, W.; Shao, A. A wheat ccaat box-binding transcription factor increases the grain yield of wheat with less fertilizer input. *Plant Physiol.* **2015**, *167*, 411–423. [CrossRef] [PubMed]

40. Zhang, L.; Zhao, G.; Jia, J.; Liu, X.; Kong, X. Molecular characterization of 60 isolated wheat MYB genes and analysis of their expression during abiotic stress. *J. Exp. Bot.* **2012**, *63*, 203–214. [CrossRef]
41. Gaufichon, L.; Reisdorfcren, M.; Rothstein, S.J. Biological functions of asparagine synthetase in plants. *Plant Sci.* **2015**, *179*, 141–153. [CrossRef]
42. Xu, G.; Fan, X.; Miller, A.J. Plant nitrogen assimilation and use efficiency. *Annu. Rev. Plant Biol.* **2012**, *63*, 153–182. [CrossRef] [PubMed]
43. Ohashi, M.; Ishiyama, K.; Kojima, S.; Konishi, N.; Nakano, K.; Kanno, K. Asparagine synthetase1, but not asparagine synthetase2, is responsible for the biosynthesis of asparagine following the supply of ammonium to rice roots. *Plant Cell Physiol.* **2015**, *56*, 769–778. [CrossRef] [PubMed]
44. Shin, R.; Berg, R.H.; Schachtman, D.P. Reactive Oxygen Species and Root Hairs in Arabidopsis Root Response to Nitrogen, Phosphorus and Potassium Deficiency. *Plant Cell Physiol.* **2005**, *46*, 1350–1357. [CrossRef]
45. Apel, K.; Hirt, H. Reactive oxygen species: Metabolism, oxidative stress, and signal transduction. *Annu. Rev. Plant Biol.* **2004**, *55*, 373–399. [CrossRef] [PubMed]
46. Lian, X.M.; Wang, S.P.; Zhang, J.W.; Feng, Q.; Zhang, L.D.; Fan, A.L. Expression profiles of 10,422 genes at early stage of low nitrogen stress in rice assayed using a cDNA microarray. *Plant Mol. Biol.* **2006**, *60*, 617–631. [CrossRef] [PubMed]
47. Wan, H.S.; Li, J.; Chen, H.H.; Wang, L.L.; Peng, Z.S.; Hu, X.R. Preliminary identification of the biologic characteristics of Chuanmai 42 under different N, P, K concentration. *Southwest China J. Agric. Sci.* **2007**, *20*, 281–285.
48. Boris, L.; Tomáš, L.; Manschadi, A.M. Arbuscular mycorrhizae modify winter wheat root morphology and alleviate phosphorus deficit stress. *Plant Soil Environ.* **2018**, *64*, 47–52. [CrossRef]
49. Qiu, Z.; Yuan, M.; He, Y.; Li, Y.; Zhang, L. Physiological and transcriptome analysis of He-Ne laser pretreated wheat seedlings in response to drought stress. *Sci. Rep.* **2017**, *7*, 6108. [CrossRef] [PubMed]
50. Langmead, B.; Trapnell, C.; Pop, M. Ultrafast and memory-efficient alignment of short DNA sequences to the human genome. *Genome Biol.* **2009**, *10*, R25. [CrossRef]
51. Trapnell, C.; Pachter, L.; Salzberg, S.L. TopHat: Discovering splice junctions with RNA-Seq. *Bioinformatics* **2009**, *25*, 1105–1111. [CrossRef]
52. Kenneth, J.; Livak, T.D. Analysis of relative gene expression data using real-time quantitative PCR and the $2^{-\Delta\Delta Ct}$ method. *Method* **2001**, *25*, 402–408.

© 2019 by the authors. Licensee MDPI, Basel, Switzerland. This article is an open access article distributed under the terms and conditions of the Creative Commons Attribution (CC BY) license (http://creativecommons.org/licenses/by/4.0/).

Article

Different Roles of Heat Shock Proteins (70 kDa) During Abiotic Stresses in Barley (*Hordeum vulgare*) Genotypes

Simone Landi [1], Giorgia Capasso [1], Fatma Ezzahra Ben Azaiez [2], Salma Jallouli [2], Sawsen Ayadi [2], Youssef Trifa [2] and Sergio Esposito [1,*]

[1] Dipartimento di Biologia, Università di Napoli "Federico II", Via Cinthia, I-80126 Napoli, Italy
[2] Department of Agronomy and Plant Biotechnology, National Institute of Agronomy, 1082 Tunis, Tunisia
* Correspondence: sergio.esposito@unina.it

Received: 25 June 2019; Accepted: 24 July 2019; Published: 26 July 2019

Abstract: In this work, the involvement of heat shock proteins (HSP70) in barley (*Hordeum vulgare*) has been studied in response to drought and salinity. Thus, 3 barley genotypes usually cultivated and/or selected in Italy, 3 Middle East/North Africa landraces and genotypes and 1 improved genotype from ICARDA have been studied to identify those varieties showing the best stress response. Preliminarily, a bioinformatic characterization of the HSP70s protein family in barley has been made by using annotated Arabidopsis protein sequences. This study identified 20 putative HSP70s orthologs in the barley genome. The construction of un-rooted phylogenetic trees showed the partition into four main branches, and multiple subcellular localizations. The enhanced HSP70s presence upon salt and drought stress was investigated by both immunoblotting and expression analyses. It is worth noting the Northern Africa landraces showed peculiar tolerance behavior versus drought and salt stresses. The drought and salinity conditions indicated the involvement of specific HSP70s to counteract abiotic stress. Particularly, the expression of cytosolic MLOC_67581, mitochondrial MLOC_50972, and encoding for HSP70 isoforms showed different expressions and occurrence upon stress. Therefore, genotypes originated in the semi-arid area of the Mediterranean area can represent an important genetic source for the improvement of commonly cultivated high-yielding varieties.

Keywords: drought; salinity; poaceae; HSP70; landraces; mediterranean area; chaperons; abiotic stress

1. Introduction

The Heat Shock Proteins 70 (HSP70s) are a subfamily of the heat shock proteins, a well-known class of molecular chaperons involved in an abiotic stress response [1]. The HSP70s present a nucleotide binding domain (NBD) of 45kDa showing ATPase activity and a 15 kDa substrate binding domain (SBD) with a C-terminal domain covering the SBD [2]. The C-term region acts as a lid and cooperate with SBD in substrates binding [3,4]. The SBD differs among the species and usually presents organelle specific motifs [5,6]. Particularly, a plants' HSP70s show several different subcellular localizations, namely cytosolic, nuclear, endoplasmic reticulum, chloroplastic and mitochondrial [7,8].

The HSP70s play a central role in the stabilization of proteins both under optimal conditions and during stress, thus helping cellular machinery in verifying protein quality and regulating protein degradation [9]. Particularly, the HSP70s avoid the aggregation of polypeptides and facilitate the proteins' maturation [10]. During abiotic stresses, the HSP70s act on misfolded and truncated proteins thus protecting the cells and the tissues [11,12]. This mechanism is regulated by heat shock factors (Hsfs), a group of transcription factors regulating HSP70s expression [12,13].

The HSP70 activation during environmental perturbations has been reported in different plants such as *Arabidopsis thaliana*, *Brachypodium distachyon*, *Glycine max*, *Capsicuum annum*, *Solanum lycopersium* and others [4,11–14]. Particularly, water scarcity and soil salinity, together with nitrogen deprivation, represent critical factors for a crops' production [15–18].

Nowadays, the improvement of crop yields in adverse environments represents one of the most impelling topics [19,20] (Tester and Langridge, 2010; Cobb et al., 2013). Particularly, the key role of Poaceae in food demand is well recognized: Rice (*Oryza sativa*), wheat (*Triticuum aestivum*), maize (*Zea mays*) and barley (*Hordeum vulgare*) represent the most important food sources for the majority of the world's population [21–24]. In this context, barley represents a critical agronomic resource in semi-desert environments, especially in the Southern Europe and Northern Africa. In developing countries, barley is a critical component of cereal rotations, playing a key role in the integrated crop-livestock production systems. It provides a stable source for sustaining smallholder farmers, replacing wheat or other cereals in many arid areas [25]. Barley shows a natural resistance to exogenous stimuli, thus representing the most tolerant Poacea against abiotic stresses [23]. Furthermore, 40% of alleles were maintained in cultivated barley compared with the historic progenitor (*Hordeum spontaneum*—[26]). This wild ancestor showed remarkable tolerance to salt, drought and heavy metals stress, but domestication by humans and, more recently, breeding programs produced high-yielding barley cultivars that are, on the other hand, more sensitive to abiotic stress, making this aspect a critical issue in barley as well [23,24,27].

Therefore, the ancestral and local cereal landraces that originated from saline and emarginated environments could represent a source of genetic diversity [25,28,29]. Therefore, the breeding research focus is moved to minimizing the gap between yields under optimal and stress conditions [25,30], contributing to the adaptation on, and contrast to, climate change [31].

To the authors' knowledge, no molecular and enzymatic studies are present about the HSP70s in barley landraces from the Mediterranean area. The goal of this research is the evaluation of the specific presence of the HSP70s isoforms under salt and drought stress in different barley genotypes. Therefore, a bioinformatic characterization of the HSP70s protein family in barley has been made, and seven barley genotypes and landraces obtained from Italy and Northern Africa were used to investigate the occurrence of the HSP70s under abiotic stress conditions.

2. Results

2.1. HSP70s Showed Peculiar Roles against Abiotic Stress in Barley

2.1.1. NaCl and PEG Effects on Barley Plants

In order to characterize the HSP70s role(s) upon salt and drought stress, this study selected the commercial variety *Hordeum vulgare* Nure. To describe a general response pattern of barley plants to salinity and water deficit, short-term severe stress conditions (10% PEG and 150 mM NaCl) were imposed to plants grown in hydroponics. The stress response was monitored using relative water content (RWC) and proline content. As described in Figure 1A, after 3 days of treatments, the barley plants showed the maximum stress effects. Particularly, a significant 21% and 24% decreased in RWC was reported after 3 days of treatments and remained stable up to 7 days. Furthermore, the proline content increased from approximately 2 to 6–7-fold change, in NaCl stressed plants. Intriguingly, the drought induced a higher proline from 3 to 12-fold change within 7d (Figure 1B).

2.1.2. Barley HSP70 Isoforms Showed Specific Occurrence upon Abiotic Stresses

The HSP70s roles upon salt and drought stress from Nure were investigated using the different occurrence of isoforms together with specific gene expression analyses.

A western blotting approach using cyt-, chl-, and mito-HSP70 antibodies showed peculiar behavior for the different HSP70 isoforms upon abiotic stresses (Figure 2). The Salinity induced a slight increase of cytosolic HSP70 occurrence after 3 h. On the contrary, chloroplastic HSP70 remained substantially unchanged in both the control and stressed plants. Mitochondrial HSP70 was barely detectable and slightly increased after 1 day of treatment.

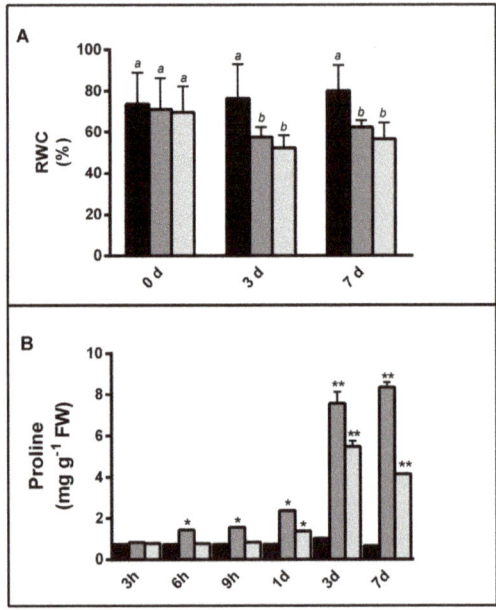

Figure 1. (**A**) The relative water content (RWC) and (**B**) Proline content, in leaves of barley (*Hordeum vulgare* cv. Nure) grown under controlled conditions (black bars); salt stress (150 mM NaCl—medium grey bars) and drought (10% PEG—light grey bars) at given times. In (**A**) statistically similar data are grouped by letters a and b in the control, salt stress and drought groups, respectively. In (**B**) asterisks indicate a significance between stressed and the control plants. * = $p < 0.05$; ** = $p < 0.001$.

The drought stress showed an increase of all HSP70 isoforms after 3 days of treatment, but cytosolic HSP70 increased soon after 9 h.

2.2. HSP70s in Barley: A Bioinformatic Overview

In order to investigate the role(s) of different HSP70s upon abiotic stress in crops, this study performed an extensive bioinformatic approach to characterize this gene family in *Hordeum vulgare*. Using annotated *Arabidopsis thaliana* protein sequences (at https://www.arabidopsis.org), the authors identified putative HSP70s orthologs in barley genome at https://ics.hutton.ac.uk/morexGenes/ and http://plants.ensembl.org/index.html, showing a total of 20 HSP70s genes (Table 1).

The intracellular localization of the HSP70 obtained by the phylogenetic analysis was confirmed by the online software Prot Comp9.0 server4 (http://linux1.softberry.com/berry.phtml), mitoproth server (http//ihg.gsf.de/ihg/mitoprot.html) and using Chloro P software (http://www.cbs.dtu.dk/services/ChloroP/)—(Table 1). Furthermore, in order to characterize the identified proteins, the pfam database was used—each protein, with the exception of MLOC_55096, retrieved the HSP70s pfam domain (PF00012.20). The prokaryotic domain PF06723.13 was retrieved in 17 protein as well. This singularity was interpreted by considering PF00012.20 and PF06723.13 belonging to the same pfam clan (CL0108).

With the aim of identifying the different HSP70 sub-families and their phylogenetic connections, a comparison of putative protein sequences was made versus *Arabidopsis thaliana* and Poaceae (Rice—*Oryza sativa* and Mais—*Zea mays*) sequences, thus obtaining an un-rooted phylogenetic tree (Figure 3a). This showed the partition into four main branches. Group 1 includes cytosolic/nuclear

HSP70s encoded by MLOC_14228, MLOC_72334 MLOC_45046, MLOC_12446, MLOC_53941, MLOC_78867 and MLOC_4447. This group of proteins showed an interesting similarity with HSP70 1-2-3-4-T1 and HSP70B from *Arabidopsis thaliana*. A second group includes the HSP70s localized within the endoplasmatic reticulum, generally recognized as BIP proteins. These HSP70s present an ATP-binding domain at the N-terminal and a C-terminal domain binding targets by recognition of the hydrophobic patches typical of improperly/incompletely folded proteins. Group 2 includes the HSP70s encoded by MLOC_77827 and MLOC_55999.

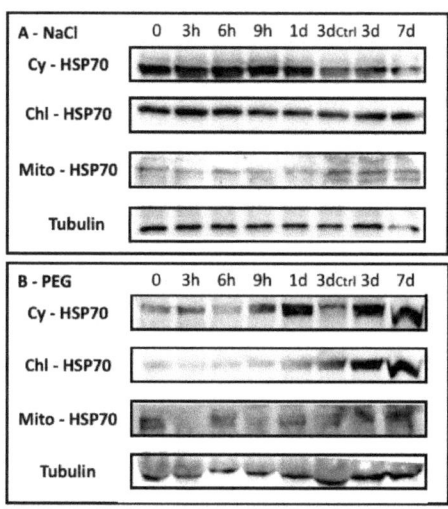

Figure 2. The immunoblotting of leaf extracts from Barley Nure plants exposed to (**A**) salt stress (NaCl); and (**B**) Drought (PEG) collected at given times. In the lane 3ctrl extracts from untreated plants after 3d were loaded. Immunoblotting was performed by using antibodies raised abainst cytosolic (Cy-HSP70); chloroplastic (Chl-HSP70) and mitochondrial (Mito-HSP70). The control blots using anti-tubulin antisera are shown.

Table 1. List of identified barley HSP70s, their localization and pfam identified domains.

Locus	Localization	Proposed Nomenclature	Pfam Domains IDs
MLOC_12446	Cytoplasm	HvHSP70	PF00012.20; PF06723.13
MLOC_14228	Cytoplasm	HvHSP70	PF00012.20; PF06723.13
MLOC_15242	Mitochondrial	HvMithHSP70	PF00012.20; PF06723.13
MLOC_2467	Cytoplasm	HvHSP70	PF00012.20; PF06723.13
MLOC_26505	Cytoplasm	HvHSP70	PF00012.20; PF06723.13
MLOC_37101	Chloroplast	HvCHPHSP70	PF00012.20; PF06723.13
MLOC_4447	Cytoplasm	HvHSP70	PF00012.20; PF06723.13
MLOC_45046	Cytoplasm	HvHSP70	PF00012.20; PF06723.13
MLOC_50972	Mitochondrial	HvMithHSP70	PF00012.20; PF06723.13; PF02782.16
MLOC_53941	Cytoplasm	HvHSP70	PF00012.20; PF06723.13
MLOC_55086	Chloroplast	HvCHPHSP70	PF00012.20; PF06723.13
MLOC_55096	Cytoplasm	HvHSP70	PF00685.27
MLOC_55999	Cytoplasm	HvBIP	PF00012.20; PF06723.13
MLOC_61727	Mitochondrial	HvMithHSP70	PF00012.20; PF06723.13
MLOC_65512	Cytoplasm	HvHSP70	PF00012.20; PF06723.13
MLOC_67581	Cytoplasm	HvHSP70	PF00012.20
MLOC_72334	Cytoplasm	HvHSP70	PF00012.20; PF06723.13
MLOC_76167	Cytoplasm	HvHSP70	PF00012.20
MLOC_77827	Cytoplasm	HvBIP	PF00012.20; PF06723.13
MLOC_78867	Cytoplasm	HvHSP70	PF00012.20; PF06723.13

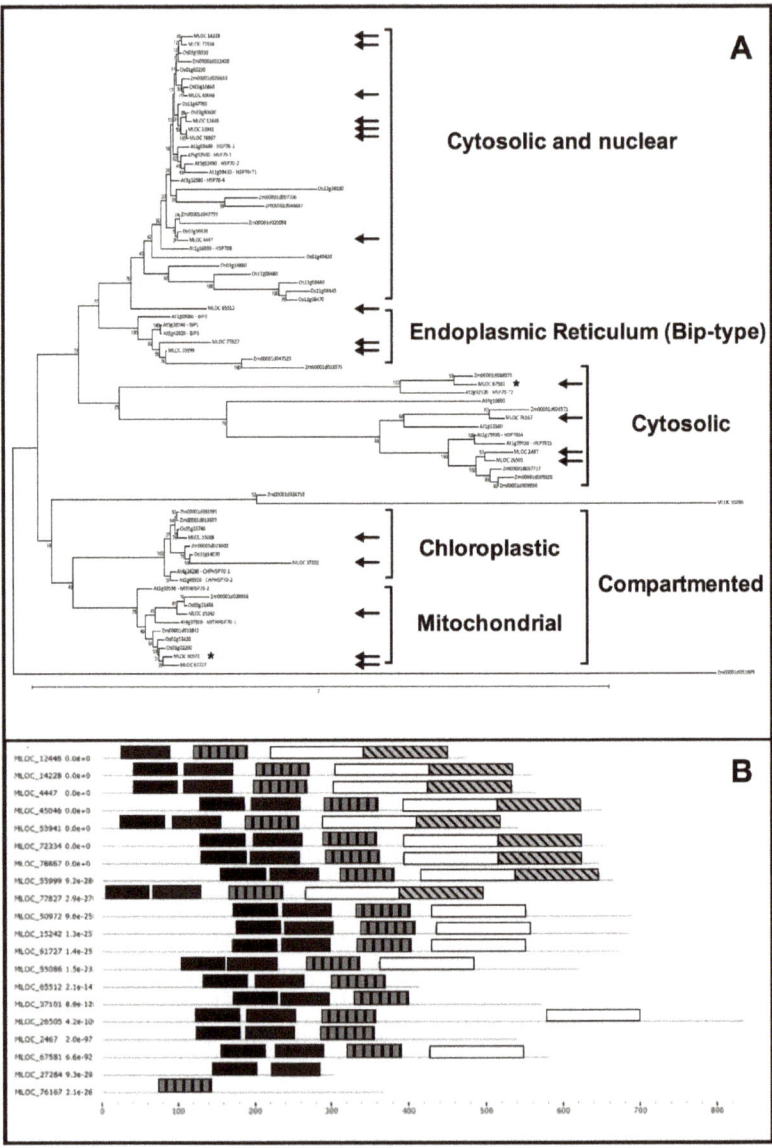

Figure 3. (**A**) A phylogenetic tree obtained by comparison of barley HSP70 amino acidic sequences of translated genes performed versus the correspondent *Arabidopsis thaliana*, *Oryza sativa* and *Zea mays* sequences. The predicted subcellular localizations are indicated for the four main branches; the arrows indicate the position of barley HSP70 isoforms. The asterisks indicate the two barley isoforms utilized for further expression studies. See text for further details. (**B**) Conserved domain analysis of barley HSP70 proteins, obtained using the MEME bioinformatic tools (http://meme-suite.org). The different color boxes represent different types of domains: ATPase binding domains (black, grey and vertical line pattern), substrate binding domain (white block) and the c-term lid (crossing line pattern). The number indicated the position of amino acids in protein sequences.

A third group includes cytoplasmic HvHSP70s similar to *Arabidopsis thaliana* HSP70-T2 (MLOC_67581) and *At*HSP7014-15 (MLOC_2467, MLOC_26505).

Further, the fourth branch identified the HSP70s present into organelles. In fact, this splits in two further forks, including chloroplastic-HSP70s (MLOC_55086 and MLOC_37101) and mitochondrial-HSP70s (MLOC_15242, MLOC_50972 and MLOC_61727). Finally, two HSP70s apparently cluster outside the major branches (MLOC_65512 and MLOC_55096).

A conserved domain analysis, using the MEME bioinformatic tools (http://meme-suite.org), was carried out to investigate the HSP70s protein structures. As showed in Figure 3B, MLOC_55096 protein showed no-HSP70 domains and probably do not represent a member of this class of proteins, therefore it was excluded in this analysis. MLOC_65512, MLOC_37101, MLOC_2467, MLOC_26505 and MLOC_76167 showed a less conserved substrate binding domain (Figure 3B).

Furthermore, compartmented HSP70s MLOC_50972, MLOC_15242, MLOC_61727, MLOC_55086 and MLOC_37101 showed no lid domain. This property is common to other cytosolic HSP70s, such as MLOC_65512, MLOC_2467, MLOC_67581 (Figure 3B).

A bioinformatic survey on *Arabidopsis* orthologous was performed to identify the best co-expressed genes (Supplemental Table S1). This verifies the cross-interaction among various HSP70s and/or other members of the heat shock proteins family (HSP20s, HSP80s). It is also worth noting the interesting relationship showed by the cytosolic HSP70 1-2-3 (*At*5g02490, *At*5g02500, *At*3g09440), and a mitochondrial HSP70 (*At*5g09590) which showed strictly a co-expression. These three cytosolic HSP70s were suggested to be participating to the abiotic stress response (Leng et al., 2017). These HSP70s showed a co-expression with stress related genes as glutathione-s-transferase (*At*5g42150), pyrroline-5-carboxylase-reductase (*At*5g14800), FTSH proteases 4 and 10 (*At*2g26140 and *At*1g07510), ascorbate peroxidase and dehydroascorbate reductase (*At*1g07890 and *At*1g75270) and others (Data not shown).

Furthermore, an expression analysis of barley HSP70s was attained by using the online RNA-seq dataset at https://ics.hutton.ac.uk/morexGenes/. As showed in Table 2, 10 HSP70s genes (MLOC_2467, MLOC_12446, MLOC_26505, MLOC_37101, MLOC_50972, MLOC_53941, MLOC_55086, MLOC_55999, MLOC_76167, MLOC_78867) are constitutively expressed in each tissue/development stage. Among these, MLOC_12446 appears the barley HSP70 predominant expressed gene. Furthermore, 4 HSP70 genes (MLOC_14228, MLOC_4447, MLOC_45046 and MLOC_67581) showed seedling and/or grains specific expression, while 4 HSP70s genes (MLOC_55096, MLOC_65512 and MLOC_77827) showed no FPKM counts in the control conditions, and these are probably regulated upon specific stimuli.

2.3. Analysis of cis-Acting Elements in the HSP70s Promoters

To investigate the regulation patterns of HSP70s, a search on the cis-elements in the promoter regions (1500 bp upstream from to the start codons) was made by using the PLANTCARE database (Table 3). Particularly, the barley HSP70 genes highlighted different behaviors to counteract the abiotic stresses. The drought sensitive elements (MBS) were found in MLOC_12446, MLOC_2467, MLOC_45046, MLOC_50972, MLOC_53941, MLOC_55096, MLOC_65512, MLOC_67581, MLOC_72334 and MLOC_78867; heat-responsive elements (HSE) were found in MLOC_12446, MLOC_2467, MLOC_26505, MLOC_4447, MLOC_45046, MLOC_50972, MLOC_55096, MLOC_55999, MLOC_61727, MLOC_65512 and MLOC_78867.

Table 2. The expression analysis of barley's HSP70 genes obtained at https://ics.hutton.ac.uk/morexGenes in different development stages and tissues: Embryogenesis, seedling shoots and roots, inflorescences (young and development), development tillers and development grains (5 and 15 days post anthesis DPA). The data are expressed as FPKM normalized counts of three different replicates. The colors represent the expression value from lower values (0—red) from higher values (green).

	Expression data (FPKM)							
	4 Days Embryo	Seedling Shoots	Young Inflorescences	Developing Inflorescences	Seedling Roots	Developing Tillers	Developing Grains (5 DPA)	Developing Grains (15 DPA)
MLOC_12446	928.19	674.36	1584.58	1909.81	1181.38	354.58	524.31	216.92
MLOC_14228	39.44	202.24	1.21	3.92	418.49	4.72	72.14	220.23
MLOC_15242	33.28	33.49	32.72	28.73	38.93	1.82	38.89	6.71
MLOC_2467	224.4	139.3	166.2	182.7	203.6	308.2	193.2	123.5
MLOC_26505	45.00	67.59	50.43	68.15	67.48	11.90	94.58	60.19
MLOC_37101	135.86	237.33	139.46	98.02	60.39	71.76	116.04	45.66
MLOC_4447	0.04	13.67	0.02	2.03	32.01	0.27	0.38	43.40
MLOC_45046	101.50	336.15	3.47	2.63	226.17	210.66	77.81	20.03
MLOC_50972	171.19	103.86	196.05	153.72	109.15	39.96	170.41	44.16
MLOC_53941	162.85	134.29	322.87	549.16	175.32	92.78	302.49	79.18
MLOC_55086	51.30	125.24	45.13	46.13	69.61	36.50	118.59	86.64
MLOC_55096	1.68	0.13	0.00	0.00	2.58	0.53	0.04	0.00
MLOC_55999	267.44	166.85	126.02	118.81	259.95	64.44	720.53	146.05
MLOC_61727	3.53	11.46	0.92	12.78	23.20	0.23	13.54	26.91
MLOC_65512	0.00	0.03	0.00	0.01	1.62	0.00	0.00	0.01
MLOC_67581	0.80	6.53	0.40	0.58	17.17	0.01	2.02	44.56
MLOC_72334	2.86	16.57	0.08	0.16	54.85	0.05	5.23	6.86
MLOC_76167	36.80	27.78	52.17	54.60	27.87	46.08	44.69	13.62
MLOC_77827	0.47	0.18	0.07	0.01	0.65	0.37	0.29	0.24
MLOC_78867	336.38	242.84	726.65	1088.08	384.78	169.08	511.81	164.32

Table 3. The regulatory cis-acting elements of the barley HSP70s promoters (1500 pb upstream region). The legend for motifs: ABRE and motif IIB (abscisic acid response); ARE (anaerobic induction); AUXRR-COR (auxin response); GCTCA & TGAGC (Me-Jasmonate—response); Box W1 & EIRE (Elicitor responsive elements); ERE (ethylene-responsive element); GARE & P-BOX (gibberellin-responsive element); GCN4 & Skn-1_motif (regulatory element required for endosperm expression); HSE (heat stress response); MBS (MYB binding site involved in drought response); LTR (in low-temperature response); p-BOX (gibberellin-responsive element); Ry elements (seed-specific regulation); TATC (gibberellin-response); TC-rich repeats (defense and stress response); TCA (salicylic acid response); TGA (auxin- response).

											Cis-Actig Elements in Promoter Region									
	ABRE	ARE	AUXRR-Core	BOX W1	CCAAT	CGTCA	ERE	EIRE	GARE	GCN4	HSE	Light	LTR	P-box	MBS	Motif IIB	Ry	SKN-1	TATC	TC-rich	TCA	TGA	TGAGC
MLOC_12446	0	1	1	0	1	0	2	0	0	0	2	8	0	0	1	0	0	3	1	3	0	0	0
MLOC_14228	1	1	0	0	0	0	0	0	1	0	0	4	0	1	0	0	0	1	0	1	0	0	0
MLOC_15242											No available data												
MLOC_2467	0	0	2	0	1	4	0	1	3	0	1	18	4	0	2	0	0	1	0	2	0	1	4
MLOC_26505	2	1	0	1	0	2	0	0	0	1	1	7	1	0	0	1	0	2	0	1	0	0	2
MLOC_37101											No available data												
MLOC_4447	0	1	0	0	0	0	0	0	0	2	2	14	0	0	0	0	0	1	0	3	2	0	0
MLOC_45046	2	2	0	1	1	1	1	0	1	0	0	17	2	0	2	0	1	2	0	0	0	0	0
MLOC_50972	0	0	0	3	0	3	0	0	1	0	1	20	0	0	4	1	0	5	0	0	1	0	1
MLOC_53941	1	1	0	0	0	1	0	0	3	0	0	10	1	1	3	0	1	5	0	1	1	0	3
MLOC_55086											No available data												
MLOC_55096	2	2	0	0	0	1	0	0	3	1	1	27	0	0	14	0	1	7	0	0	0	0	1
MLOC_55999	3	0	0	2	0	2	0	0	1	0	2	17	1	0	0	0	0	2	0	1	0	0	2
MLOC_61727	0	0	0	1	0	4	0	0	0	0	1	18	0	1	0	0	0	3	0	0	0	1	4
MLOC_65512	2	2	0	1	2	1	0	0	1	0	1	11	1	0	1	1	1	4	1	0	1	0	1
MLOC_67381	4	2	0	3	2	2	0	1	1	0	0	19	2	1	1	1	2	8	0	0	1	0	2
MLOC_72334	0	0	0	0	2	2	0	0	3	1	0	15	0	0	1	0	0	4	0	2	1	1	2
MLOC_76167											No available data												
MLOC_77827											No available data												
MLOC_78867	3	0	0	0	0	2	0	0	0	0	1	15	0	0	2	0	0	3	0	0	1	2	2

In addition, TC-rich repeats motif (cis-acting element related to defense and stress response) were identified in MLOC_12446, MLOC_14228, MLOC_2467, MLOC_26505, MLOC_4447, MLOC_50972, MLOC_53941, MLOC_55999, MLOC_65512 and MLOC_72334.

Furthermore, the HSP70s genes from barley exhibiting different patterns of cis acting elements in response to plant phytoregulators, such as abscissic Acid (ABRE elements and IIB motif), Gibberelic Acid (GARE elements and Box P), Auxin (TGA elements), Ethylene (ERE elements) and Methyl Jasmonate (CGTCA and TGAGC motifs).

Interestingly, MLOC_67581 presents 4 ABRE elements (responsive to ABA), the highest number among all barley HSP70s, suggesting an effective abiotic stress induction of this cytosolic isoform. On the opposite, MLOC_12446, MLOC_2467, MLOC_4447, MLOC_50972, MLOC_61727, MLOC_65512, MLOC_72334 present no ABRE elements. Among the latter group (no-ABRE elements), MLOC_50972 interestingly shows 3 BOX W1 (fungal elicitor), and 3 CGTCA elements (Me-Jasmonate responsiveness) indicating this mitochondrial isoform as strictly specific versus a pathogen attack. It should be noted that also MLOC_72334 (3 BOX W1, 2 CGTCA elements), and MLOC_61727 (1 BOX W1, 4 CGTCA elements) are suspected to be highly sensitive to fungi/pathogens.

2.4. Real Time PCR of Selected HSP70 Isoforms

The bioinformatic analysis of promoter regions allowed the identification of at least two HSP70 isoforms that appear differently regulated upon stress. As previously described, MLOC_67581 encodes for a cytosolic isoform that is characterized by the highest presence of ABA responsive elements (4 ABRE). On the other hand, the biotic stress related elements are strongly limited in the promoter of this gene.

In contrast, the mitochondrial MLOC_50972 does not present ABRE (and ARE) elements, but this promoter region shows the highest number of elements responsive to biotic stress: 3 Box W1, 3 CGTA (MeJA responsive) and 1 TC-rich elements, thus suggesting that this HSP70 isoform is induced under fungal/pathogen attack, but scarcely reactive to abiotic stress.

Therefore, this study performed a qRT-PCR expression analysis of these two genes to investigate their possible different expression rates upon abiotic stress.

As showed in Figure 4, barley plants showed a consistently increased expression of cytosolic HSP70s (MLOC_67581) both upon NaCl (over 37-fold change) and drought (over 23-fold change) compared with the control.

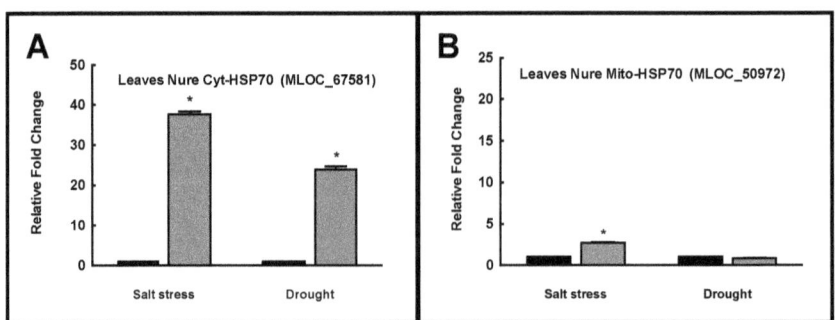

Figure 4. (**A**) The changes in the expression of Cyt-HSP70 (MLOC_67581) (abiotic stress responsive); and (**B**) Mito-HSP70 (MLOC_50972) (biotic stress responsive) in leaves of barley plants cv. Nure collected after 3 days of exposure to salt stress (NaCl 150 mM) and drought (PEG 10%). Asterisks (*) indicate p value ≤ 0.001.

The mitochondrial HSP70(MLOC_50972) showed a slight 2.5 increase of expression upon salinity while no significant differences were reported upon drought.

2.5. Effects of Abiotic Stress in in Different Barley Genotypes

The different barley genotypes and landraces were exposed to salinity and drought. When fresh weight was measured, salt stress did not induce severe changes in all genotypes, except Icarda 20 (−25%). A general and significant decrease was measured upon drought, with Nure (−13%) as the most resistant, and Icarda 20 and Batinì (−32–39%, respectively) as the most susceptible varieties (Figure 5A).

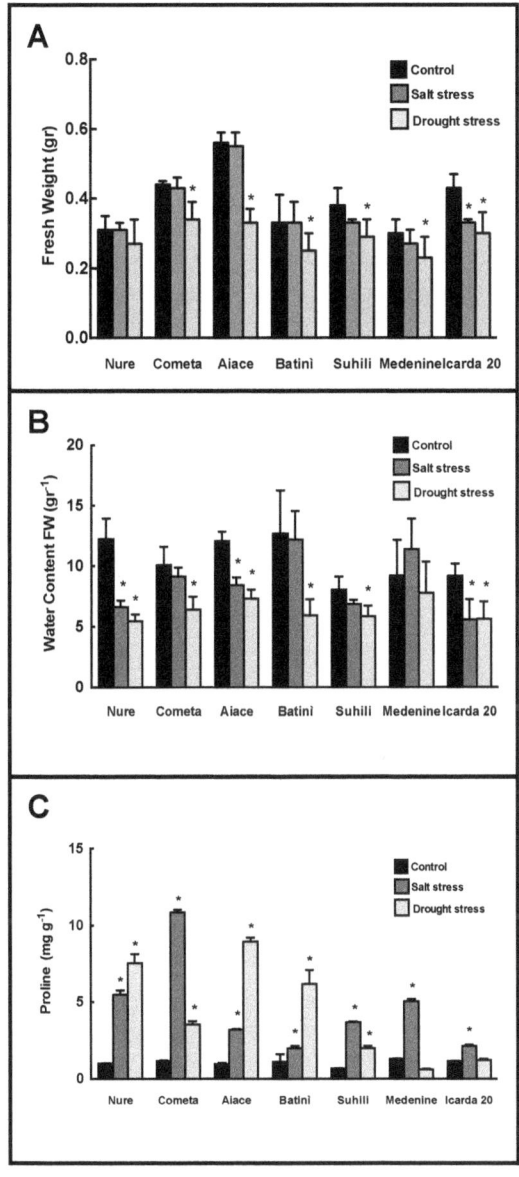

Figure 5. The changes in (**A**) Fresh weight; (**B**) Relative water content (RWC) and (**C**) Proline levels, in leaves of selected barley genotypes in control condition (black bars), Salt stress (NaCl 150 mM) (dark grey bars); and drought (PEG 10%) (light grey bars). The asteriks (*) indicate p value ≤ 0.001.

The relative water content consistently decreased in almost all varieties upon both salinity and drought. Cometa, Batinì, Suhili and Medenine were unaffected by salinity and only Medenine did not exhibit significant changes upon drought (Figure 5B).

These results were evidently counteracted by an increase in Proline. The highest increase was observed in Cometa under salt stress, and Aiace under drought. In Medenine and Icarda 20, the proline increase was among the lowest under salinity, and not significant under drought (Figure 5C).

Generally, abiotic stress induced severe effects in Italian genotypes, while Northern Africa landraces showed peculiar responses to abiotic stress. Icarda 20 appears less susceptible to treatment. The salt stress response was lower in Batinì, Suihili and Medenine.

2.6. HSP70s in Different Barley Genotypes

The HSP70s isoforms occurrence was further analyzed in barley genotypes from Italy and Northern Africa.

The western blotting analysis indicates that the cytosolic HSP70s display specific occurrence depending on stress treatments (salt or drought). Particularly, all selected genotypes showed an increase of cyt-HSP70 occurrence upon salinity. In contrast, the chl-HSP70 showed no, or reduced, changes in abiotic stress treatments among the various genotypes/landraces.

The Mito-HSP70 protein occurrence increased in the Icarda 20 genotype under stress, particularly if compared with the Italian genotypes. No appreciable changes were reported in mito-HSP70 protein for Batinì, Suihili and Medenine landraces (Figure 6).

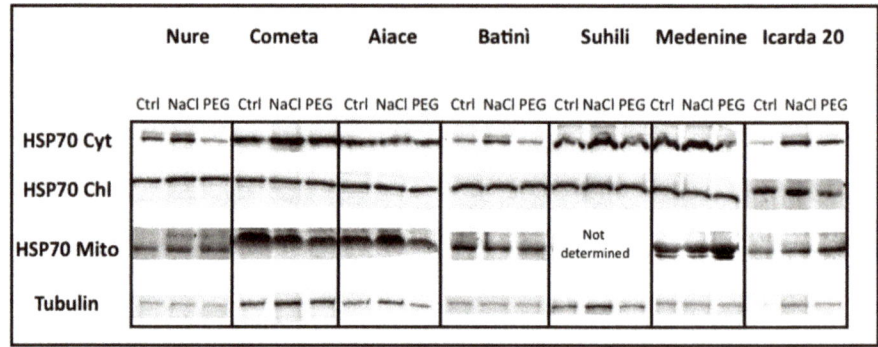

Figure 6. Immunoblotting of leaf extracts from different barley genotypes and landraces plants under control conditions (Ctrl) or exposed to salt stress (NaCl); or drought (PEG) collected after three days. Immunoblotting was performed by using antibodies raised abainst cytosolic (Cy-HSP70); chloroplastic (Chl-HSP70) and mitochondrial (Mito-HSP70). The control blots using anti-tubulin antisera are shown.

Given these results, Batinì landrace and Icarda 20 genotype were selected for a comparison with the model specie Nure through qRT-PCR analysis in the previously selected HSP70 isoforms encoded by MLOC_67581 (cytosolic, induced by abiotic stress) and MLOC_50972 (mitochondrial, sensitive to pathogen attack). Preliminarily, landraces showed a higher constitutive expression levels of cytosolic MLOC_67581, Batinì 3.4-fold, and Icarda 20 7,7-fold higher with respect to barley Nure (Supplemental Table S2).

As shown in Figure 7A, this higher constitutive level of MLOC_67581 (cytosolic) resulted in a low increase in its expression under salinity, and drought (only in Icarda 20). Batinì landrace showed a strong increase, over 50-fold of this isoform under drought.

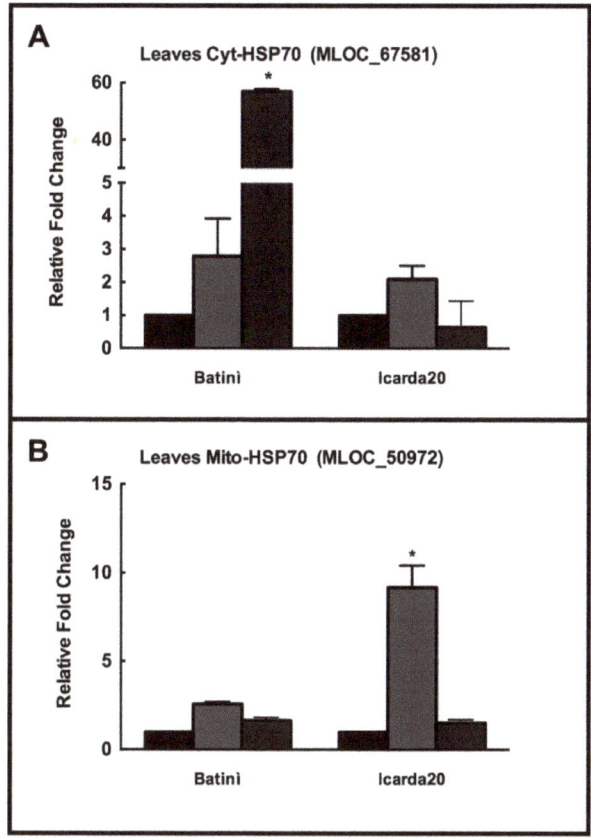

Figure 7. (**A**) The changes in the expression of Cyt-HSP70 (MLOC_67581) (abiotic stress responsive); and (**B**) Mito-HSP70 (MLOC_50972) (biotic stress responsive), in leaves of barley plants cv. Batinì and ICARDA 20 collected after 3 days upon control (Black bars), salt stress (NaCl 150 mM—light grey bars) and drought (PEG 10%—dark grey bars). The asterisks (*) indicate p value ≤ 0.001.

In contrast, mitochondrial HSP70 MLOC_50972 was substantially expressed at very similar levels (Batinì 1.2-fold, and Icarda 20, 0.56-fold with respect to barley Nure) under control conditions (Supplemental Table S2). Batinì showed no change in the expression of the mitochondrial, biotic-stress inducible MLOC_50972 under both salinity and drought. Icarda 20 showed an appreciable increase in the expression of this mitochondrial HSP70 only under salinity (about 10-fold) (Figure 7B).

3. Discussion

Barley ranks tenth as the most important produced crop worldwide, with a global cultivation estimated at approximately 143 million tons (FAO stats, 2013; http://faostat3.fao.org/browse/rankings/countries_by_commodity).

Particularly, 30% of the global barley production is targeted for malting, while 70% to feed use [32]. Traditionally, barley is mainly used as a food crop for human nutrition in the semi-arid countries of Africa (e.g., Morocco, Algeria, and Tunisia), Middle East, the Andean countries of South America and in some Asian counties (e.g., Nepal and Tibet). In European countries such as Germany, France, UK, Denmark and Italy, barley is primarily used for feeding animals [33]. An increased value was given by the managing of the brewing by-products which are feedstock for thermochemical conversion, biogas and ethanol production and other applications [34].

In recent years, climate change has reduced the European average production of barley by 3.8% because of the temperature and precipitation changes [35]. This evidence together to the commercial value of barley highlight the need of select new genotypes with improved tolerance to abiotic stress as a strategy to guarantee sustainability [36].

This work provided evidence for the contribution of specific HSP70 isoforms in plant responses to different abiotic stresses, namely drought and salinity.

Recently, different studies described the central role of HSP70s in plants in stress-response conditions [11,12,37]. The drought and salinity conditions used in this work clearly indicated the involvement of selected HSP70 isoforms to counteract the related stresses in barley.

When barley plants of the cultivar Nure were cultivated in vitro under controlled conditions, and exposed to salinity and drought, a decrease in RWC, and a concomitant increase in proline levels were observed, indicating the effectiveness of the stress imposed.

The plant response to this stress induced a differential occurrence of distinct HSP70 isoforms—cytosolic HSP70s rapidly increased, particularly upon salinity, but a long-lasting increase was observed upon drought. On the other hand, chloroplastic isoforms remained substantially unaffected under salt stress and increased upon drought conditions, while mitochondrial HSP70s increased under both stresses.

These results pose questions about the identification of distinct HSP70 isoforms induced by stress, and the conditions inducing their specific expression. Therefore, an extensive bioinformatic analysis on barley genome allowed the identification of 20 genes encoding for barley HSP70s, and their localization, the specific tissue expression and the stages of development. Among these, 16 genes are actively expressed in specific tissues and/or specific developmental stages. Similar numbers of HSP70 genes have been recently described in Arabidopsis thaliana [7], pepper [38], rice [6], poplar and *Physchomitrella patens* [39].

A quantitative RT-PCR analysis confirmed that the cytosolic isoform strongly increased its expression level upon abiotic stress, while the mitochondrial HSP70 was slightly affected only upon salinity and insensitive to drought. Similar results were showed upon abiotic stresses in other crops as tomato [24,40], pepper [38], rice [41] and others. This identifies key HSP70 genes related to stress tolerance (*Solyc*09g075950, *Solyc*03g117630, *Ca*03g30260 (*Ca*Hsp70-2) and LOC_*Os*08g39140). Intriguingly, analogous roles were identified for the HSP70s to counteract toxic effects of heavy metals in barley upon cadmium stress [37], highlighting the role of this gene family to counteract the effects of unfavorable environments.

It is therefore clear that the specific response by the HSP70s would greatly ameliorate the adaptation of specific barley cultivars and landraces under abiotic stress conditions.

Barley HSP70s present, as it could be easily assumed, different and multiple cis-acting elements in their promoter regions—cis-elements related to ABA, drought, salinity and other stresses were found in the promoters of the HSP70 genes [13,39].

Thus, two isoforms were identified that were supposed to exhibit opposite regulation upon stress. These two HSP70 isoforms are strongly suspected to undergo opposite regulation: The cytosolic HSP70 MLOC_67581, showing the highest number of ABA responsive elements and possibly under abiotic stress control; a mitochondrial isoform, presenting multiple elements involved in fungal/pathogen attack response—HSP70 MLOC_50972—thought to be inducible under pathogen attack. An expression analysis confirmed that in barley Nure, a sensible increase in MLOC_67581 was observed under drought and salinity, while MLOC_50972 was only slightly affected by abiotic stress.

Recently, the detrimental effects of modern breeding and plant domestication were reported to decrease the genomic biodiversity and reduce the abiotic stress tolerance of cultivated crops [42]. The exploitation of landraces and wild relatives is a promising strategy to counteract the genetic erosion [24,42–44].

In a second set of experiments, this study analyzed the role(s) of HSP70 isoforms in six barley varieties other than Nure: 2 italian genotypes; Aiace and Cometa; one genotype selected by ICARDA (Icarda 20); three barley genotypes and landraces from Tunisia (Suhili, Medenine) and Oman (Batinì). Interestingly, the selected Northern Africa landraces showed peculiar tolerance behavior versus drought and salt stresses. Particularly, the specific protein occurrence and gene expression increases were reported for the HSP70s as well as the proline accumulations.

Similar opportunities were recently available using barley varieties from northern Asia. Tibetan barley genome (*Hordeum vulgare* L. var. *nudum*), showed a remarkable enlargement in stress-related gene families [45]. Furthermore, Tibetan wild barley (*Hordeum spontaneum* C.) was deeply characterized because of an increased tolerance to salinity and drought obtained by a more efficient sugar and glycine-betaine accumulation, Na^+/K^+ ratio regulation, ROS detoxification and others [23,27].

These results highlighted the prospective genotypes originated from the semi-arid area of the Mediterranean as a genic source for the improvement of the high-yielding varieties. Among the six varieties investigated, the landrace Batinì showed a different response to salinity, and the improved genotype Icarda 20 resulted as less influenced by both stresses, when compared to their changes in FrWt, RWC and proline levels. These results were substantially confirmed by an immunoblotting analysis on the HSP70s occurrence.

Following these results, the expression analysis was repeated on the two test HSP70 MLOC_67581 and MLOC_50972 on the landrace Batinì and the selected genotype Icarda 20. Interestingly, both presented an enhanced expression of cytosolic HSP70 MLOC_67581 with respect to Nure. Furthermore, the level of expression did not change upon salinity. Only in the landrace Batinì was a strong enhancement of expression observed under drought conditions. The mitochondrial HSP70 MLOC_50972 did not change, except for a 10-fold increase under salinity in Batinì. These results clearly indicate that the traditional selection of landraces and the modern selection with advanced crossing techniques converge on common molecular traits—in the case here studied, the constitutive overexpression of a stress related cytosolic HsSP70 (MLOC_67581). It is intriguing that in the landrace Batinì the mitochondrial MLOC_50972 is expressed consistently upon salinity, while in the barley genome this promoter does not present cis-acting elements devoted to this stress response. It could be argued that landraces may present changes in both promoter regions of specific stress responding genes. However, the signaling cascade may be changed to adapt to the specific environment.

These evidences strongly encourage further efforts to identify abiotic stress tolerance alleles of landraces from extreme environments.

Further studies are necessary to characterize agronomical, physiological and molecular traits of the Northern Africa landraces in different experimental environments. The HSP70 genes from these genotypes could be sequenced and the genomic peculiarities of these genes and of the regulation region can be identified.

4. Materials and Methods

4.1. Plant Material and Stress Treatments

The seeds of Italian barley varieties (*Hordeum vulgare*, var. Nure, Cometa and Aiace) were supplied by Centro di ricerca per la genomica e la postgenomica animale e vegetale (CRA-GPG—Fiorenzuola D'Arda—PC, Italy). The seeds of MENA (Middle East North Africa) barley (*Hordeum vulgare*, var. Batinì, Suhili, Medenine and Icarda 20) were supplied by the Laboratory of Genetics and Cereal breeding—INAT, University of Tunis. The genotype's features were listed in Supplemental Table S3. The seeds were germinated for 7 days in the dark on moistened paper. Then, seedlings were grown in hydroponic solution in darkened plastic bottles at 20 °C, at 60–80% relative humidity, under 16h-light/8h-dark regime, with approximately 180 µmol photons m-2 s-1. The growth medium (modified Hoagland solution) was described in [46]. The solution was continuously aerated.

After 7d in hydroponics, the plants were separated in three groups: The controls were maintained in the standard solution; the salt stress was imposed by adding 150 mM of NaCl to the standard solution; the drought was imposed by the presence of 10% PEG 8000 MW, (Sigma-Aldrich), added to the hydroponic solution. The growth medium was daily controlled for volume and pH and adjusted accordingly. The leaves from Nure genotype were collected at 0 h, 3 h, 6 h, 9 h, 1 day, 3 days and 7 days after the stress induction. The leaves from the other genotypes were collected after 3 days from stress induction.

4.2. Growth Variation and Water Content Determination

The changes in the relative water content (RWC) in barley plants exposed to salt and PEG were measured at 0, 3 and 7 days after the stress imposition on 15–20 plants. The plants' weight was evaluated after hydroponic growth for FW determination. The plants were hydrated for 2–3 h by either floating in a Petri dish in distilled water and weighed to determine the turgid weight (TW). Then, the samples were dried overnight at 70 °C for dried weight (DW) measurements. The plant's RWC was calculated as follows: RWC % = (FW − DW) / (TW − DW) × 100 [47].

4.3. Proline Content

Proline was measured as in [48]. The powdered leaves (250 mg) were suspended in 1.5 mL of 3% sulphosalicylic acid, filtered through a glass-fiber filter (Macherey-Nagel, Ø 55 mm, Germany). Further, 1ml of glacial acetic acid and 1 mL ninhydrin reagent (2.5 g ninhydrin/100 mL of a 6:3:1 solution of glacial acetic acid, deionized water and 85% orto-phosphoric acid) were added to the filtrate (1 mL). After 1 h at 100 °C, the optical density was read at 546 nm (Cary 60 spectrophotometer—Agilent Technologies, Santa Clara, CA, USA).

4.4. Western Blotting

In immunoblottings, the proteins were extracted as previously described and separated by SDS-PAGE [49]. Then, the polypeptides were transferred onto a Hybond nitrocellulose membrane (GE Healthcare, Chicago, IL, USA). The filter was incubated with primary antibodies (Agrisera) versus the HSP70s (Cytosolic, Chloroplastic and mitochondrial) and tubulin. After incubation of the membrane with secondary antibodies, the cross-reacting polypeptides were identified by enhanced chemioluminescence (WesternBright™ Quantum kit—Advansta, San Josè, CA, USA). The images were acquired by BioRad Chemidoc system (Bio-Rad, Hercules CA, USA).

4.5. RNA Extraction and qRT-PCR

The RNA extraction was made from the leaves (100 mg) using Bio-Rad Aurum™ Total RNA Mini Kit. The cDNA syntheses were done using the ThermoScript RT-PCR System. The RNA amount was measured by NanoDrop ND-1000 spectrophotometer. The gene expression analysis was carried out by qRT-PCR. Triplicate quantitative assays were made by using Applied Biosystems™ 7500 Real-Time PCR System and Platinum SYBR Green qPCR SuperMix (Life Technologies, Carlsbad, CA, USA). The leaf samples of the control plants were used as calibrators and α-tubulin served as an endogenous reference gene. The quantization of the gene expression was carried out using the $2^{-\Delta\Delta Ct}$ method as in [50]. the mRNA amount was calculated in each sample, relative to the calibrator sample for the same gene. A list of primers is provided in Supplemental Table S4.

4.6. Bioinformatics Analysis

The sequences of barley HSP70s were found using barley genome at https://ics.hutton.ac.uk/morexGenes/. The sequences from different species were from the TAIR database (https://www.arabidopsis.org) and the Ensamble plants database. The alignments and phylogenetic analyses were made by the software MEGA 6.0 [51]. The alignments were obtained by MUSCLE algorithm.

The phylogenetic tree was designed by using the maximum likelihood method with the JTT substitution model. The test of phylogeny was performed using a bootstrap method (bootstrap replication = 100). The conserved motif analysis was performed by MEMESuite4.11.1 server 5 [52]. The promoter analyses were performed at Plant CARE server suites using regions of 1000 bp upstream from the start codons of each HSP70 gene [53]. An *Arabidopsis thaliana* orthologs co-expression analysis was carried out by ATTED-II versus 8.0 (http://atted.jp). The degree of co-expression was estimated as mutual rank [54]. The expression analysis in different tissues was retrieved using the database at https://ics.hutton.ac.uk/morexGenes/.

4.7. Statistics

The experiments were made in at least three replicates. The values were expressed as the mean ± standard error and statistical through the Student's t-test ($p \leq 0.05$). The ANOVA analysis was used to compute the statistical significance of differences between the controls and the stressed groups and between different genotypes (ANOVA corresponds to $\alpha = 0.05$). The Tukey–Kramer post-hoc test was used to evaluate differences between the means.

Supplementary Materials: The following are available online at http://www.mdpi.com/2223-7747/8/8/248/s1, Table S1: Co-expression analysis of Arabidopsis HSP70S orthologous, obtained using the ATTED-II database, Table S2: Expression rate of MLOC_67581 and MLOC_50972 by qRT-PCR in barley Batinì and Icarda 20 vs. Barley Nure, Table S3: List of selected genotypes of barley (*Hordeum vulgare*) utilized in this study, Table S4: List of primers used for qRT-PCR analysis.

Author Contributions: Conceptualization, S.L., G.C. and S.E.; methodology, plant and seed material S.E., Y.T., S.A., F.E.B.A.; investigation, G.C., F.E.B.A., S.L.; resources, Y.T., S.E.; writing—original draft preparation, S.L., G.C.; writing—review, S.J., S.L.; editing and supervision, S.E.

Funding: Research supported by Legge Regionale della Campania 5/2002 (2007), CUP E69D15000270002 to S.E.

Acknowledgments: The Authors wish to thank CRA-GPG—Fiorenzuola D'Arda—PC, Italy and Laboratory of Genetics and Cereal breeding—INAT, University of Tunis for the generous gift of barley seeds.

Conflicts of Interest: The authors declare no conflicts of interest.

References

1. Sung, D.Y.; Vierling, E.; Guy, C.L. Comprehensive expression profile analysis of the Arabidopsis Hsp70 gene family. *Plant Physiol.* **2001**, *126*, 789–800. [CrossRef]
2. Yu, A.; Li, P.; Tang, T.; Wang, J.; Chen, Y.; Liu, L. Roles of Hsp70s in Stress Responses of Microorganisms, Plants, and Animals. *BioMed Res. Int.* **2015**, *2015*, 510319. [CrossRef] [PubMed]
3. Dragovic, Z.; Broadley, S.A.; Shomura, Y.; Bracher, A.; Hartl, F.U. Molecular chaperones of the Hsp110 family act as nucleotide exchange factors of HSP70s. *EMBO J.* **2006**, *25*, 2519–2528. [CrossRef] [PubMed]
4. Zhang, L.; Zhao, H.K.; Dong, Q.L.; Zhang, Y.Y.; Wang, Y.M.; Li, H.Y.; Xing, G.J.; Li, Q.Y.; Dong, Y.S. Genome-wide analysis and expression profiling under heat and drought treatments of HSP70 gene family in soybean (Glycine max L.). *Front. Plant Sci.* **2015**, *6*, 773. [CrossRef] [PubMed]
5. Zhu, X.; Zhao, X.; Burkholder, W.F.; Gragerov, A.; Ogata, C.M.; Gottesma, M.E.; Hendrickson, W.A. Structural analysis of substrate binding by the molecular chaperone DnaK. *Science* **1996**, *272*, 1606–1614. [CrossRef] [PubMed]
6. Sarkar, N.K.; Kundnani, P.; Grover, A. Functional analysis of Hsp70 superfamily proteins of rice (Oryza sativa). *Cell Stress Chaperon* **2013**, *18*, 427–437. [CrossRef] [PubMed]
7. Lin, B.L.; Wang, J.S.; Liu, H.C.; Chen, R.W.; Meyer, Y.; Barakat, A.; Delseny, M. Genomic analysis of the Hsp70 superfamily in Arabidopsis thaliana. *Cell Stress Chaperon* **2001**, *6*, 201–208. [CrossRef]
8. Usman, M.G.; Rafii, M.Y.; Martini, M.T.; Yusuff, O.A.; Ismail, R.M.; Miah, G. Molecular analysis of Hsp70 mechanisms in plants and their function in response to stress. *Biotechnol. Genet. Eng. Rev.* **2017**, *33*, 26–39. [CrossRef]
9. Hartl, F.U.; Bracher, A.; Hayer-Hartl, M. Molecular chaperones in protein folding and proteostasis. *Nature* **2011**, *475*, 324–332. [CrossRef]

10. Su, P.H.; Li, H.M. Arabidopsis stromal 70-kD heat shock proteins are essential for plant development and important for thermotolerance of germinating seeds. *Plant Physiol.* **2008**, *146*, 1231–1241. [CrossRef]
11. Leng, L.; Liang, Q.; Jiang, J.; Zhang, C.; Hao, Y.; Wang, X.; Su, W. A subclass of HSP70s regulate development and abiotic stress responses in Arabidopsis thaliana. *J. Plant Res.* **2017**, *130*, 349–363. [CrossRef] [PubMed]
12. Wen, F.; Wu, X.; Li, T.; Jia, M.; Liu, X.; Li, P.; Zhou, X.; Ji, X.; Yue, X. Genome-wide survey of heat shock factors and heat shock protein 70s and their regulatory network under abiotic stresses in Brachypodium distachyon. *PLoS ONE* **2017**, *12*, e0180352. [CrossRef] [PubMed]
13. Guo, M.; Lu, J.P.; Zhai, Y.F.; Chai, W.; Gong, Z.H.; Lu, M.H. Genome-wide analysis, expression profile of heat shock factor gene family (CaHsfs) and characterisation of CaHsfA2 in pepper (*Capsicum annuum* L.). *BMC Plant Biol.* **2015**, *15*, 151. [CrossRef] [PubMed]
14. Landi, S.; Nurcato, R.; De Lillo, A.; Lentini, M.; Grillo, S.; Esposito, S. Glucose-6-phosphate dehydrogenase plays a central role in the response of tomato (Solanum lycopersicum) plants to short and long-term drought. *Plant Physiol. Biochem.* **2016**, *105*, 79–89. [CrossRef] [PubMed]
15. Hayashi, H.; Murata, N. Genetically engineered enhancement of salt tolerance in higher plants. In *Stress Response of Photosynthetic Organisms: Molecular Mechanisms and Molecular Regulation*; Sato, K., Murata, N., Eds.; Elsevier: Amsterdam, The Netherlands, 1998; pp. 133–148.
16. Reynolds, M.; Tuberosa, R. Translational research impacting on crop productivity in drought-prone environments. *Curr. Opin. Plant Biol.* **2008**, *11*, 171–179. [CrossRef] [PubMed]
17. Curci, P.L.; Aiese Cigliano, R.; Zuluaga, D.L.; Janni, M.; Sanseverino, W.; Sonnante, G. Transcriptomic response of durum wheat to nitrogen starvation. *Sci. Rep.* **2017**, *26*, 1176. [CrossRef] [PubMed]
18. Landi, S.; Esposito, S. Nitrate uptake affects cell wall synthesis and modelling. *Front. Plant Sci.* **2017**, *8*, 1376. [CrossRef] [PubMed]
19. Tester, M.; Langridge, P. Breeding technologies to increase crop production in a changing world. *Science* **2010**, *12*, 818–822. [CrossRef] [PubMed]
20. Cobb, J.N.; Declerck, G.; Greenberg, A.; Clark, R.; McCouch, S. Next-generation phenotyping: Requirements and strategies for enhancing our understanding of genotype-phenotype relationships and its relevance to crop improvement. *Theor. Appl. Genet.* **2013**, *126*, 867–887. [CrossRef]
21. Gill, B.S.; Appels, R.; Botha-Oberholster, A.M.; Buell, C.R.; Bennetzen, J.L.; Chalhoub, B.; Chumley, F.; Dvorák, J.; Iwanaga, M.; Keller, B.; et al. A workshop report on wheat genome sequencing: International Genome Research on Wheat Consortium. *Genetics* **2004**, *168*, 1087–1096. [CrossRef]
22. Cui, P.; Liu, H.; Islam, F.; Li, L.; Farooq, M.A.; Ruan, S.; Zhou, W. OsPEX11, a Peroxisomal Biogenesis Factor 11, Contributes to Salt Stress Tolerance in Oryza sativa. *Front. Plant Sci.* **2016**, *7*, 1357. [CrossRef] [PubMed]
23. Shen, Q.; Fu, L.; Dai, F.; Jiang, L.; Zhang, G.; Wu, D. Multi-omics analysis reveals molecular mechanisms of shoot adaption to salt stress in Tibetan wild barley. *BMC Genom.* **2016**, *17*, 889. [CrossRef] [PubMed]
24. Landi, S.; Hausman, J.F.; Guerriero, G.; Esposito, S. Poaceae vs. abiotic stress: Focus on drought and salt stress, recent insight and perspectives. *Front. Plant Sci.* **2017**, *8*, 1214. [CrossRef] [PubMed]
25. Hammami, Z.; Gauffreteau, A.; BelhajFraj, M.; Sahli, A.; Jeuffroy, M.H.; Rezgui, S.; Bergaoui, K.; McDonnell, R.; Trifa, Y. Predicting yield reduction in improved barley (Hordeum vulgare L.) varieties and landraces under salinity using selected tolerance traits. *Field Crops Res.* **2017**, *211*, 10–18. [CrossRef]
26. Ellis, R.P.; Forster, B.P.; Robinson, D.; Handley, L.L.; Gordon, D.C.; Russell, J.R.; Powell, W. Wild barley: A source of genes for crop improvement in the 21st century? *J. Exp. Bot.* **2000**, *51*, 9–17. [CrossRef]
27. Ahmed, I.M.; Dai, H.; Zheng, W.; Cao, F.; Zhang, G.; Sun, D.; Wu, F. Genotypic differences in physiological characteristics in the tolerance to drought and salinity combined stress between Tibetan wild and cultivated barley. *Plant Physiol. Biochem.* **2013**, *63*, 49–60. [CrossRef] [PubMed]
28. Ayadi, S.; Karmous, C.; Chamekh, Z.; Hammami, Z.; Baraket, M.; Esposito, S.; Rezgui, S.; Trifa, Y. Effects of nitrogen rates on response of grain yield and nitrogen agronomic efficiency (NAE) of durum wheat cultivars under different environments. *Ann. Appl. Biol.* **2016**, *168*, 264–273. [CrossRef]
29. Bouhaouel, I.; Gfeller, A.; Boudabbous, K.H.; Fauconnier, M.L.; Slim Amara, H.; du Jardin, P. Physiological and biochemical parameters: New tools to screen barley root exudate allelopathic potential (Hordeum vulgare L. subsp. vulgare). *Acta Physiol. Plant* **2018**, *40*, 38. [CrossRef]
30. Cattivelli, L.; Rizza, F.; Badeck, W.; Mazzucotelli, E.; Mastrangelo, A.M.; Francia, E.; Marè, C.; Tondellia, A.; Stanca, M. Drought tolerance improvement in crop plants: An integrated view from breeding to genomics. *Field Crops Res.* **2008**, *105*, 1–14. [CrossRef]

31. Challinor, A.J.; Watson, J.; Lobell, D.J.; Howden, S.M.; Smith, D.; Chetri, N. A meta-analysis of crop yield under climate change and adaptation. *Nat. Clim. Chang.* **2014**, *4*, 287–291. [CrossRef]
32. Akar, T.; Avci, M.; Dusunceli, F. *Barley: Post-Harvest Operations*; FAO in PHO Post Harvest Compendium: Ankara, Turkey, 2004.
33. Bhatty, R.S. Hulless barley: Development and Civilisation. In *VII International Barley Genetics Symposium*; University of Saskatoon: Saskatoon, SK, Canada, 1986; pp. 106–111.
34. Mussatto, S.I. Brewer's spent grain: A valuable feedstock for industrial applications. *J. Sci. Food Agric.* **2014**, *94*, 1264–1275. [CrossRef] [PubMed]
35. Moore, F.C.; Lobell, D.B. The fingerprint of climate trends on European crop yields. *Proc. Nalt. Acad. Sci. USA* **2015**, *112*, 2670–2675. [CrossRef] [PubMed]
36. Boyer, J.S. Drought decision-making. *J. Exp. Bot.* **2010**, *61*, 3493–3497. [CrossRef] [PubMed]
37. Lentini, M.; De Lillo, A.; Paradisone, V.; Liberti, D.; Landi, S.; Esposito, S. Early responses to cadmium exposure in barley plants: Effects on biometric and physiological parameters. *Acta Physiol. Plant* **2018**, *40*, 178. [CrossRef]
38. Guo, M.; Liu, J.H.; Ma, X.; Zhai, Y.F.; Gong, Z.H.; Lu, M.H. Genome-wide analysis of the Hsp70 family genes in pepper (Capsicum annuum L.) and functional identification of CaHsp70-2 involvement in heat stress. *Plant Sci.* **2016**, *252*, 246–256. [CrossRef] [PubMed]
39. Tang, T.; Yu, A.; Li, P.; Yang, H.; Liu, G.; Liu, Li. Sequence analysis of the Hsp70 family in moss and evaluation of their functions in abiotic stress responses. *Sci. Rep.* **2016**, *6*, 33650. [CrossRef] [PubMed]
40. Iovieno, P.; Punzo, P.; Guida, G.; Mistretta, C.; Van Oosten, M.J.; Nurcato, R.; Bostan, H.; Colantuono, C.; Costa, A.; Bagnaresi, P.; et al. Transcriptomic changes drive physiological responses to progressive drought stress and rehydration in tomato. *Front. Plant Sci.* **2011**, *7*, 371. [CrossRef] [PubMed]
41. Ouyang, S.; Zhu, W.; Hamilton, J.; Lin, H.; Campbell, M.; Childs, K.; Thibaud-Nissen, F.; Malek, R.L.; Lee, Y.; Zheng, L.; et al. The TIGR rice genome annotation resource: Improvements and new features. *Nucleic Acids Res.* **2007**, *35*, D846–D851. [CrossRef]
42. Dwivedi, S.L.; Ceccarelli, S.; Blair, M.W.; Upadhyaya, H.D.; Are, A.K.; Ortiz, R. Landrace Germplasm for Improving Yield and Abiotic Stress Adaptation. *Trends Plant Sci.* **2016**, *21*, 31–42. [CrossRef]
43. Lopes, S.M.; El-Basyoni, I.; Baenziger, P.S.; Singh, S.; Royo, C.; Ozbek, K.; Aktas, H.; Ozer, E.; Ozdemir, F.; Manickavelu, A.; et al. Exploiting genetic diversity from landraces in wheat breeding for adaptation to climate change. *J. Exp. Bot.* **2015**, *66*, 3477–3486. [CrossRef]
44. Landi, S.; De Lillo, A.; Nurcato, R.; Grillo, S.; Esposito, S. In-field study on traditional Italian tomato landraces: The constitutive activation of the ROS scavenging machinery reduces effects of drought stress. *Plant Physiol. Biochem.* **2017**, *118*, 150–160. [CrossRef] [PubMed]
45. Zeng, X.; Long, H.; Wang, Z.; Zhao, S.; Tang, Y.; Huang, Z.; Wang, Y.; Xu, Q.; Mao, L.; Deng, G.; et al. The draft genome of Tibetan hulless barley reveals adaptive patterns to the high stressful Tibetan Plateau. *Proc. Nalt. Acad. Sci. USA* **2015**, *112*, 1095–1100. [CrossRef] [PubMed]
46. Cardi, M.; Castiglia, D.; Ferrara, M.; Guerriero, G.; Chiurazzi, M.; Esposito, S. The effects of salt stress cause a diversion of basal metabolism in barley roots: Possible different roles for glucose-6-phosphate dehydrogenase isoforms. *Plant Physiol. Biochem.* **2015**, *86*, 44–54. [CrossRef] [PubMed]
47. Matin, M.; Brown, J.H.; Ferguson, H. Leaf water potential, relative water content, and diffusive resistance as screening techniques for drought resistance in barley. *Agron. J.* **1989**, *81*, 100–105. [CrossRef]
48. Claussen, W. Proline as a measure of stress in tomato plants. *Plant Sci.* **2005**, *168*, 241–248. [CrossRef]
49. Castiglia, D.; Cardi, M.; Landi, S.; Cafasso, D.; Esposito, S. Expression and characterization of a cytosolic glucose-6-phosphate dehydrogenase isoform from barley (Hordeum vulgare) roots. *Protein Express Purif.* **2015**, *112*, 8–14. [CrossRef]
50. Livak, K.J.; Schmittgen, T.D. Analysis of relative gene expression data using real-time quantitative PCR and the 2-$\Delta\Delta$CT method. *Methods* **2001**, *25*, 402–408. [CrossRef]
51. Tamura, K.; Stecher, G.; Peterson, D.; Filipski, A.; Kumar, S. MEGA6: Molecular Evolutionary Genetics Analysis version 6.0. *Mol. Biol. Evol.* **2013**, *30*, 2725–2729. [CrossRef]
52. Bailey, T.L.; Bodén, M.; Buske, F.A.; Frith, M.; Grant, C.E.; Clementi, L.; Ren, J.; Li, W.W.; Noble, W. MEME SUITE: Tools for motif discovery and searching. *Nucleic Acids Res.* **2009**, *37*, W202–W208. [CrossRef]

53. Lescot, M.; Dehais, P.; Thijs, G.; Marchal, K.; Moreau, Y.; Van de Peer, Y.; Rouze, P.; Rombauts, S. Plantcare, a data base of plant cis-acting regulatory elements and a portal to tools for in silico analysis of promoter sequences. *Nucleic Acids Res.* **2002**, *30*, 325–327. [CrossRef]
54. Aoki, Y.; Okamura, Y.; Tadaka, S.; Kinoshita, K.; Obayashi, T. ATTED-II in 2016: A plant co-expression database towards lineage-specific co-expression. *Plant Cell Physiol.* **2016**, *57*, e5. [CrossRef] [PubMed]

© 2019 by the authors. Licensee MDPI, Basel, Switzerland. This article is an open access article distributed under the terms and conditions of the Creative Commons Attribution (CC BY) license (http://creativecommons.org/licenses/by/4.0/).

Article

Comparative Transcriptome Analysis of Waterlogging-Sensitive and Tolerant Zombi Pea (*Vigna vexillata*) Reveals Energy Conservation and Root Plasticity Controlling Waterlogging Tolerance

Pimprapai Butsayawarapat [1], Piyada Juntawong [1,2,3,*], Ornusa Khamsuk [4] and Prakit Somta [5]

1. Department of Genetics, Faculty of Science, Kasetsart University, Bangkok 10900, Thailand
2. Center for Advanced Studies in Tropical Natural Resources, National Research University-Kasetsart University, Bangkok 10900, Thailand
3. Omics Center for Agriculture, Bioresources, Food and Health, Kasetsart University (OmiKU), Bangkok 10900, Thailand
4. Department of Botany, Faculty of Science, Kasetsart University, Bangkok 10900, Thailand
5. Department of Agronomy, Faculty of Agriculture at Kamphaeng Saen, Kasetsart University, Nakhon Pathom 73140, Thailand
* Correspondence: fscipdj@ku.ac.th; Tel.: +66-02-562-5555

Received: 22 June 2019; Accepted: 31 July 2019; Published: 2 August 2019

Abstract: *Vigna vexillata* (zombi pea) is an underutilized legume crop considered to be a potential gene source in breeding for abiotic stress tolerance. This study focuses on the molecular characterization of mechanisms controlling waterlogging tolerance using two zombi pea varieties with contrasting waterlogging tolerance. Morphological examination revealed that in contrast to the sensitive variety, the tolerant variety was able to grow, maintain chlorophyll, form lateral roots, and develop aerenchyma in hypocotyl and taproots under waterlogging. To find the mechanism controlling waterlogging tolerance in zombi pea, comparative transcriptome analysis was performed using roots subjected to short-term waterlogging. Functional analysis indicated that glycolysis and fermentative genes were strongly upregulated in the sensitive variety, but not in the tolerant one. In contrast, the genes involved in auxin-regulated lateral root initiation and formation were expressed only in the tolerant variety. In addition, cell wall modification, aquaporin, and peroxidase genes were highly induced in the tolerant variety under waterlogging. Our findings suggest that energy management and root plasticity play important roles in mitigating the impact of waterlogging in zombi pea. The basic knowledge obtained from this study can be used in the molecular breeding of waterlogging-tolerant legume crops in the future.

Keywords: *De novo* transcriptome; lateral root; legume; *Vigna vexillata*; waterlogging

1. Introduction

Flooding is one of the most significant problems facing global agriculture today. It can be categorized as waterlogging, i.e., when the height of the water column covers only the root-zone, or as submergence, when the aerial plant tissues are fully covered [1]. Waterlogging generally affects dryland crops rather than submergence, since soil can easily become waterlogged due to poor drainage after intensive and/or extensive rainfall or irrigation. Waterlogging creates low oxygen environments in the root, due to the limited diffusion of oxygen and other gases under water. This results in ATP shortage from the inhibition of oxidative phosphorylation, and long-term waterlogging results in stomatal closure, leading to impaired root hydraulic conductivity and reduced photosynthesis and nutrient and water uptake in the plants [2].

The characterization of the molecular mechanisms for submergence tolerance has been extensively studied in model plants. Functional characterization of group VII of ethylene response factor (ERF) genes revealed their functional role as critical players regulating submergence tolerance in rice and *Arabidopsis* [3–5]. In rice, natural genetic variations of group VII ERFs determine the escape strategy through stem elongation in the deepwater rice and the quiescence strategy through the restriction of shoot elongation in the lowland rice [6,7]. In another monocot model, *Brachypodium distachyon*, transcriptomic analysis of submergence-tolerant and sensitive natural genetic variations revealed the oxidative stress pathway as a significant tolerance factor [8]. Most of the submergence and low oxygen studies in model plants provide some basic understanding; however, these studies were frequently conducted in a hypoxic environment under complete darkness, which cannot imitate the impact observed in plant response to waterlogging [2].

The legume family (*Fabaceae*) is one of the most important food crops for human nutritional needs. However, molecular characterization of the mechanisms controlling waterlogging tolerance in the legume family is uneven. Most existing studies on the molecular basis of waterlogging tolerance in legumse were focused on soybean. A key component associated with waterlogging stress in soybean is an energy crisis in root-zone resulting from low oxygen conditions, with the meristem showing particular susceptibility. Waterlogging-tolerant soybean varieties were found to develop more aerenchyma and promote more root growth than the sensitive varieties under waterlogging stress [9–12]. The natural genetic diversity of soybean has been used to find molecular mechanisms that are differentially expressed in tolerant versus sensitive varieties [13–17]. Recently, a major quantitative trait locus (QTL), *qWT_Gm_03*, controlling root plasticity under waterlogging was identified in soybean and proposed to be involved in the auxin pathways regulating secondary root development and root plasticity [17]. In other legume species, higher root porosity and the ability to form lateral roots was also correlated with waterlogging tolerance, as observed in the waterlogging-tolerant legume of the genus *Trifolium* [18], pea (*Pisum sativum*) [19], and lentil (*Lens culinaris*) [19].

The genus *Vigna* is a particularly important legume crop, comprising more than 1000 species and distributed in extensive and diverse areas of Africa, America, Australia, and Asia [20,21]. Domesticated *Vigna* species including cowpea (*V. unguiculata*), zombi pea (*V. vexillata*), Bambara groundnut (*V. subterranean*), mungbean (*V. radiata*), azuki bean (*V. angularis*), rice bean (*V. umbellata*), black gram (*V. mungo*), moth bean (*V. aconitifolia*), and créole bean (*V. reflexo-pilosa*) are grown mainly for dry seeds by small farmers in several cropping systems of tropical and sub-tropical regions [22,23]. Most of the domesticated *Vigna* species are particularly sensitive to waterlogging, resulting in poor seed quality and significant yield reduction. In the case of mungbean, waterlogging at the vegetative stage results in decreased leaf area, growth rate, root growth, photosynthesis rate, chlorophyll and carotenoid contents, flowering rate, pod setting, yield, and altered dry matter partitioning [24]. In contrast to soybean, little is known about the molecular mechanisms of waterlogging tolerance in the genus *Vigna*. Therefore, to improve *Vigna* waterlogging tolerance, mechanisms of waterlogging tolerance must be understood. It has been proposed that stress-resistant plant species closely related to the crop of interest could be used for the molecular analysis of the stress adaptation mechanisms [25]. Thus, de novo transcriptome analysis and gene expression profiling can be used to provide a basic understanding of the molecular response controlling waterlogging adaptation of the non-model *Vigna* crops.

Vigna vexillata (common name: zombi pea) is an underutilized legume crop that can be found in diverse areas of Africa, America, Australia, and Asia [26]. It is cultivated for edible storage roots and seeds. Zombi pea is a highly heterogeneous legume species [21]. Previous research has found that some varieties of zombi pea adapt well to environmental stresses including infertile soil [27], alkaline soil [28], drought [29], and waterlogging [30]. Therefore, zombi pea is considered to be a potential gene source in breeding for tolerance to abiotic stresses.

In this work, we investigated the changes in anatomy, morphology, and molecular responses to waterlogging with the assistance of RNA-sequencing (RNA-seq) of the waterlogged roots of two zombi pea varieties with contrasting waterlogging-tolerant phenotypes. We hypothesized that the natural genetic diversity of the zombi pea would allow us to find the molecular mechanism of waterlogging tolerance in the genus *Vigna*.

2. Results and Discussion

2.1. Anatomical and Morphological Changes of Zombi Pea Varieties Subjected to Waterlogging Stress

Two zombi pea varieties, the waterlogging-tolerant "A408" and the waterlogging-sensitive "Bali", were selected based on the contrasting phenotype in response to waterlogging. "A408", a native pasture on the verge of a swamp, is highly waterlogging-tolerant [30]. "Bali" is an Asian cultivated zombi pea found in Bali, Indonesia [21]. The contrasting phenotype of these natural varieties was initially tested by growing them in pot soils and waterlogging for 30 days (data not shown). In this study, we applied waterlogging at the seedlings stage (15 day-old). Under long-term waterlogging stress (WS), "A408" was able to maintain growth based on the visual examination (Figure 1A). On the other hand, "Bali" displayed stunted growth of its phenotype under WS (Figure 1B). Analysis of leaf chlorophyll content demonstrated that under non-stress (NS), "A408" maintained its chlorophyll content between 0.014 mg/cm^2 at day zero and 0.017 mg/cm^2 at day seven (Figure 1C). Similarly, "Bali" chlorophyll contents ranged between 0.014 mg/cm^2 at day zero and 0.018 mg/cm^2 at day seven under NS (Figure 1D). Further, WS did not affect "A408" chlorophyll content (Figure 1C). In contrast, the reduction of "Bali" chlorophyll content was observed between day four (0.01 mg/cm^2) and day seven (0.008 mg/cm^2) of WS (Figure 1D), suggesting that WS resulted in a decline in "Bali" leaf photosynthesis. The decrease in leaf chlorophyll content under WS was also observed in other WS-sensitive legume varieties [19,31].

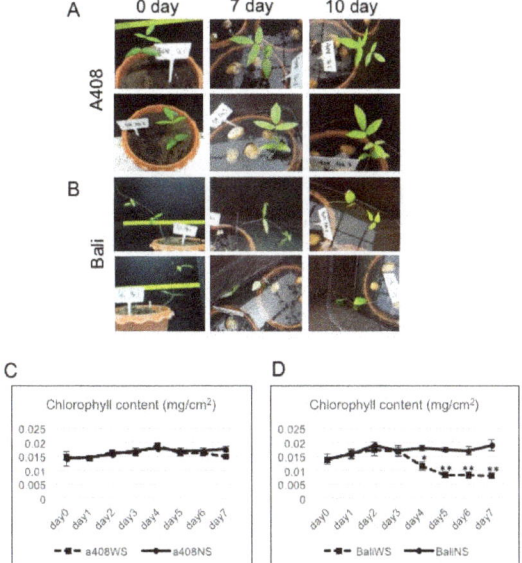

Figure 1. Contrasting waterlogging tolerance in "A408" and "Bali" varieties. Representative 15-day-old zombi pea seedlings subjected to 0, 7, and 10 days of waterlogging. (**A**) "A408". (**B**) "Bali". Leaf chlorophyll measurement (*n* = 12 plants) under no stress (NS) and waterlogging (WS). (**C**) "A408". (**D**) "Bali". * $p < 0.05$, ** $p < 0.01$ (*t*-test).

Root architecture and plasticity play a vital role in the adaptation of plants to WS [32]. Therefore, we analyzed for WS-induced root anatomical and morphological changes in zombi pea and found that

WS caused damage and significantly suppressed the root growth of "Bali" (Figure 2A,B). On the other hand, WS promoted basal stem thickening and lateral root production in "A408" (Figure 2A,B). Lateral roots are all roots that emerge from main roots, and are a major determinant of root architecture, which is essential for the efficient uptake of water and nutrients [33]. To determine the taproot anatomy, roots at the same age were sectioned at almost the same position (Figure 3A). In the cutting area, we observed the secondary root growth in "A408" but the primary root growth in "Bali". We randomly cut "Bali" roots in different root zones, but only the primary root growth was observed. Based on "A408" anatomy, "A408" taproots functions as storage roots. Therefore, the cortical region of A408 taproot was smaller than that of "Bali". Furthermore, we observed numerous starch grains in the parenchyma of "A408" root steles. In addition, WS resulted in the formation of aerenchyma in taproots and hypocotyls of "A408" variety (Figure 3A,B, respectively). In contrast, WS caused severe tissue damage in taproot of "Bali", as observed by dark precipitation of Fast Green dye, and no aerenchyma was observed in WS hypocotyls of "Bali" (Figure 3A,B). The formation of aerenchyma was responsible for increasing internal oxygen diffusion from the aerial parts to the waterlogged roots, which allowed the underground roots to maintain aerobic respiration [34]. Our results correlated with the three previous studies in other legume crops. First, in waterlogging-tolerant legumes of the genus *Trifolium*, higher root porosity and the ability to form lateral roots contributed to waterlogging tolerance [18]. Second, in waterlogging-tolerant pea and lentil, WS increased both the main and lateral root porosity compared to the NS due to the formation of aerenchyma [19]. Lastly, a soybean locus, *qWT_Gm_03*, enhanced waterlogging tolerance through controlling secondary root growth in a waterlogging-tolerant cultivar [17]. Since lateral root formation was induced by WS in "A408", but root growth was arrested in WS "Bali" (Figure 2A,B), these results suggest that the plasticity in lateral root development under WS could be an important determinant for waterlogging tolerance in zombi pea.

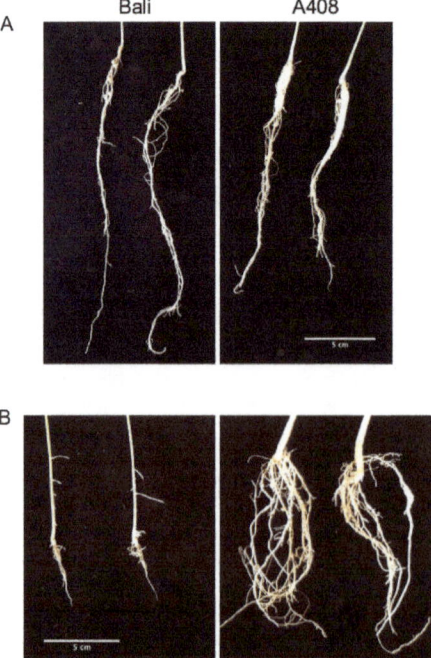

Figure 2. Changes of zombi pea root architecture under WS. (**A**) Roots of control plants kept for 7 days under NS. (**B**). Roots of 7-day WS plants.

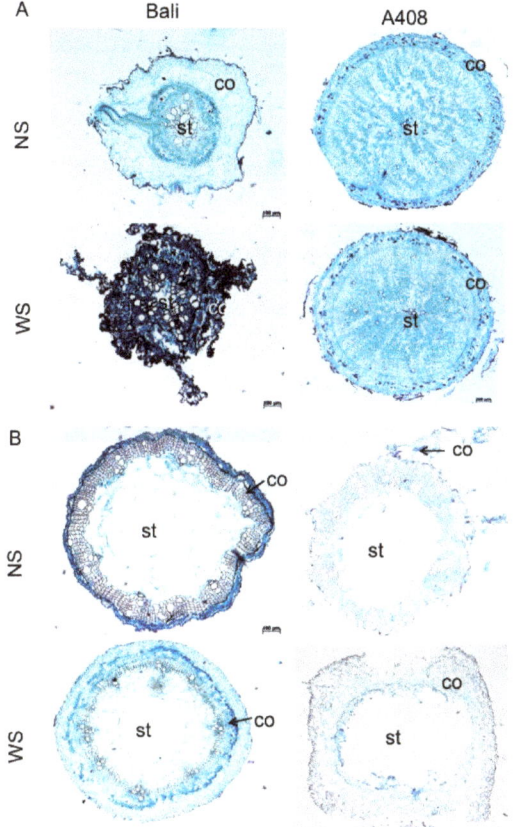

Figure 3. Waterlogging induces aerenchyma and extra-cellular airspace in hypocotyls and roots of "A408". Cross-section of "A408" and "Bali" (**A**) taproot and (**B**) hypocotyl. co = cortex. st = stele.

2.2. De Novo Transcriptome Analysis

To examine the molecular mechanisms controlling waterlogging tolerance in zombi pea, we performed de novo transcriptome analysis by RNA-seq using WS and NS root samples derived from both "A408" and "Bali" varieties. Twenty-two to twenty-six million reads were obtained for each RNA-seq library (Table S1). To construct a reference transcriptome for each variety, the RNA-seq reads from four libraries (two biological replicates per each treatment) were combined and subjected to de novo transcriptome assembly by Trinity program. The transcriptome assembly yielded 74,658 genes consist of 154,405 transcripts with an average transcript length of 1263 bp (N50 = 2134 bp) and GC content of 39.70%, and 80,279 genes consisting of 173,848 transcripts with an average transcript length of 1230 bp (N50 = 2087 bp) and GC content of 39.74% for "A408" and "Bali", respectively (Files S1 and S2; Table S1). The de novo transcriptome assembly statistics were similar between the two varieties.

To perform functional characterization of the de novo assembled transcriptomes, the candidate open reading frames of each transcript (>100 amino acids; 94,801 and 106,142 protein-coding transcripts from "A408" and "Bali", respectively) were annotated using BLASTP to plant UniprotPK database to obtain the associated gene ontology (GO) terms and assigned to functional bins by Mercator pipeline (Table S2). Transcript homologs among "A408", "Bali", and *Arabidopsis* were identified by the OrthoVenn2 web tool. Transcript expression, as represented by count per million (CPM) expression values can be found in Table S3.

2.3. Differential Gene Expression, Functional Enrichment, and Comparative Transcriptome Analyses

For differential gene expression analysis, reads were mapped back to the assembled transcriptome. The majority of reads (96–97%) from each RNA-seq library could be mapped to the reference transcriptome (Table S1), suggesting the reliability of our transcriptome data. The number of reads aligned back to each transcript was acquired for differential gene expression analysis. Transcriptome analysis identified 982 differentially-expressed genes (DEGs) and 1133 DEGs with significant changes in gene expression evaluated by the false discovery rate (FDR) <0.05 from "A408" and "Bali", respectively (Figure 4A; Table S3). For "Bali", 51% and 49% of DEGs were upregulated and downregulated by WS (Figure 4A; Table S3). On the other hand, a higher percentage of DEGs (61%) were downregulated compared to the percentage of upregulated DEGs (39%) in "A408" (Figure 4A, Table S3).

Using a list of core hypoxia-induced genes in *Arabidopsis* [35], we were able to identify 31 core hypoxia homolog clusters among "A408", "Bali", and *Arabidopsis* (Table S3). Of these, four homolog clusters, including *sucrose synthase* (cluster 56), *alcohol dehydrogenase* (cluster 3967), *similar to RCD one 5* (*SRO5*; cluster 4428), and *wound-responsive family protein* (cluster 8884), were induced in both "A408" and "Bali" (Table S3). *Non-symbiotic hemoglobin 1* (cluster 13294) was induced only in "A408". In contrast, *1-aminocyclopropane-1-carboxylate oxidase 1* (*ACC oxidase 1*; cluster 15158), *haloacid dehalogenase-like hydrolase (HAD) superfamily protein* (cluster 6574), and *LOB domain-containing protein 41* (*LBD41*, cluster 9883) were specifically induced in "Bali".

We took two contemporary approaches to identify differentially-expressed molecular mechanisms controlling waterlogging tolerance; GO enrichment analysis of co-expressed genes (Figure 4B) and comparative transcriptome analysis (Figure 5). To obtain a global picture of transcriptome adjustment in response to WS, GO enrichment analysis was carried out. The top five GO terms of upregulated DEGs of "A408" were enriched for protein unfolding, response to hydrogen peroxide, chloroplast thylakoid membrane, water transmembrane transporter activity, and asparagine biosynthetic process (Figure 4B). On the other hand, the top five GO terms of upregulated DEGs of "Bali" were enriched in response to decreased oxygen levels, cytosol, response to hydrogen peroxide, glycolytic process, and response to temperature stimulus (Figure 4B). Response to hydrogen peroxide, protein phosphatase inhibitor activity, and alcohol dehydrogenase (NAD) activity were the common GO terms that were identified from the upregulated DEGs of both "A408" and "Bali" (Figure 4B). The top five downregulated DEGs of "A408" were enriched for naringenin-chalcone synthase activity, chalcone biosynthetic process, positive regulation of post-embryonic development, sulfur compound biosynthetic process, and maltose biosynthetic process (Figure 4B). In "Bali", the top five downregulated DEGs were enriched in guanosine deaminase activity, phosphoenol pyruvate carboxykinase activity, carbohydrate derivative catabolic process, zinc ion transport, serine-type carboxypeptidase activity, farnesyltranstransferase activity, terpene synthase activity, gibberellin 3-beta-dioxygenase activity, indole acetic acid carboxyl (IAA) methyltransferase activity, (-)-secoisolariciresinol dehydrogenase activity, and 3-hydroxybutyrate dehydrogenase activity (Figure 4B).

To compare the changes in WS transcriptome in the two zombi pea varieties with contrasting WS responses, comparative transcriptome analysis was analyzed by over-representation analysis (ORA) using Fisher's exact test with a cut-off of two. The results from the ORA analysis demonstrated that glycolysis, stress, MYB-related transcription factor family, and protein functional bins were overrepresented in the upregulated DEGs of "Bali" (Figure 5; Table S4). In contrast, the upregulated DEGs of "A408" were overrepresented with cell wall, peroxidase, MYB-related transcription factor family, AUX/IAA transcription factor family, and cytoskeleton functional bins (Figure 5; Table S4). The downregulated DEGs of "Bali" were overrepresented with secondary metabolism, hormone metabolism (including gibberellin), and transport functional bins (Figure 5; Table S4). For "A408", the downregulated DEGs were overrepresented with lipid metabolism, WRKY transcription factor family, and signaling functional bins (Figure 5; Table S4). The results from the GO enrichment and the ORA analyses point out that differential regulation of the genes encoding for energy production pathways, hormones,

RNA-regulation by AUX/IAA family, cell wall modification, water transmembrane transporters, and peroxidase enzymes could contribute to waterlogging tolerance in zombi pea.

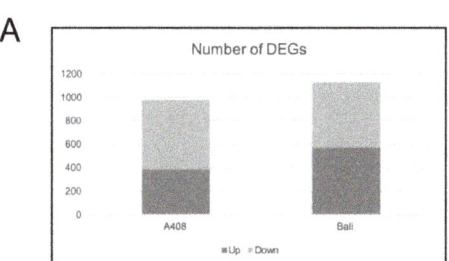

	GO ID	GO Term	log10 adjusted P-values			
			A408_Up	Bali_Up	A408_Down	Bali_Down
	GO:0043335	protein unfolding	-3.44			
	GO:0042542	response to hydrogen peroxide	-3.41	-5.44		
	GO:0009535	chloroplast thylakoid membrane	-3.06			
	GO:0005372	water transmembrane transporter activity	-2.17			
	GO:0006529	asparagine biosynthetic process	-2.13			
	GO:0009644	response to high light intensity	-1.97			
	GO:0048529	magnesium-protoporphyrin IX monomethyl ester (oxidative) cyclase activity	-1.97			
	GO:0043617	cellular response to sucrose starvation	-1.71			
	GO:0009044	xylan 1,4-beta-xylosidase activity	-1.67			
	GO:0009055	electron transfer activity	-1.67			
	GO:0010242	oxygen evolving activity	-1.67			
	GO:0004864	protein phosphatase inhibitor activity	-1.50	-1.87		
	GO:0045727	positive regulation of translation	-1.44			
	GO:0004022	alcohol dehydrogenase (NAD) activity	-1.37	-4.16		
	GO:0015200	methylammonium transmembrane transporter activity	-1.37			
	GO:0036293	response to decreased oxygen levels		-7.52		
	GO:0005829	cytosol		-6.81		
	GO:0006096	glycolytic process		-4.58		
	GO:0009266	response to temperature stimulus		-4.36		
	GO:0009744	response to sucrose		-3.77		
	GO:0046686	response to cadmium ion		-3.67		
	GO:0009413	response to flooding		-3.41		
	GO:0032355	response to estradiol		-3.41		
	GO:0008886	glyceraldehyde-3-phosphate dehydrogenase (NADP+) (non-phosphorylating) activity		-3.41		
	GO:0031000	response to caffeine		-3.36		
	GO:0048046	apoplast		-2.82		
	GO:0003979	UDP-glucose 6-dehydrogenase activity		-1.87		
	GO:0046983	protein dimerization activity		-1.78		
	GO:0051775	response to redox state		-1.64		
	GO:0004332	fructose-bisphosphate aldolase activity		-1.64		
	GO:0005618	cell wall		-1.35		
	GO:0006065	UDP-glucuronate biosynthetic process		-1.32		
	GO:0006094	gluconeogenesis		-1.32		
	GO:0016210	naringenin-chalcone synthase activity			-4.61	
	GO:0009715	chalcone biosynthetic process			-3.32	
	GO:0048582	positive regulation of post-embryonic development			-1.98	
	GO:0044272	sulfur compound biosynthetic process			-1.71	
	GO:0000024	maltose biosynthetic process			-1.61	
	GO:0003858	3-hydroxybutyrate dehydrogenase activity			-1.61	-1.97
	GO:0008242	omega peptidase activity			-1.31	
	GO:0010222	stem vascular tissue pattern formation			-1.31	
	GO:0047974	guanosine deaminase activity				-3.44
	GO:0004612	phosphoenolpyruvate carboxykinase (ATP) activity				-2.55
	GO:1901136	carbohydrate derivative catabolic process				-2.51
	GO:0006829	zinc ion transport				-2.34
	GO:0004185	serine-type carboxypeptidase activity				-1.97
	GO:0004311	farnesyltranstransferase activity				-1.97
	GO:0010333	terpene synthase activity				-1.97
	GO:0016707	gibberellin 3-beta-dioxygenase activity				-1.97
	GO:0051749	indole acetic acid carboxyl methyltransferase activity				-1.97
	GO:0102911	(-)-secoisolariciresinol dehydrogenase activity				-1.97
	GO:0010369	chromocenter				-1.51
	GO:0004568	chitinase activity				-1.50
	GO:0008308	voltage-gated anion channel activity				-1.50
	GO:0018685	alkane 1-monooxygenase activity				-1.50
	GO:0043693	monoterpene biosynthetic process				-1.31

Figure 4. Waterlogging altered root transcriptomes of "A408" and "Bali". (**A**) The number of upregulated and downregulated differentially-expressed genes (DEGs) from roots of "A408" and "Bali" in response to WS. (**B**) Enrichment of GO terms from upregulated and downregulated DEGs from roots of "A408" and "Bali" in response to WS.

Bin ID	Bin name	Up-regulation Bali	Up-regulation A408	Down-regulation Bali	Down-regulation A408
4	glycolysis	4.2	0.0	-4.2	0.0
10	cell wall	0.0	3.0	0.0	-3.0
11	lipid metabolism	0.0	-3.6	0.0	3.7
16	secondary metabolism	-3.9	0.0	3.9	0.0
17	hormone metabolism	-2.9	0.0	2.9	0.0
17.6	hormone metabolism.gibberelin	-3.9	0.0	3.9	0.0
20	stress	3.1	0.0	-3.1	0.0
26.12	misc.peroxidases	0.0	2.6	0.0	-2.6
27.3.26	RNA.regulation of transcription.MYB-related transcription factor family	2.4	2.3	-2.4	-2.3
27.3.32	RNA.regulation of transcription.WRKY domain transcription factor family	0.0	-2.8	0.0	2.8
27.3.40	RNA.regulation of transcription.Aux/IAA family	0.0	2.6	0.0	-2.6
29	protein	4.7	0.0	-4.7	0.0
30	signalling	0.0	-3.7	0.0	3.7
31.1.1	cell.organisation.cytoskeleton	0.0	2.9	0.0	-2.9
34	transport	-3.2	0.0	3.2	0.0

Figure 5. Comparative transcriptome response for selected functional categories to WS in roots of "A408" and "Bali". Over-representation analysis of the DEGs (FDR < 0.05). The statistical analysis of overrepresented functional categories was performed using Fisher method. Z-scores indicate over/under representation. (Number indicates z-score; Yellow indicates over-representation). Data used to generate this figure can be found in Table S4.

2.4. Waterlogging Resulted in Stronger Induction of Glycolysis and Fermentative Genes in "Bali" than in "A408"

Since WS creates a low oxygen environment that could promote glycolysis and fermentation and the GO enrichment and ORA analyses suggested differential expressions of glycolysis and fermentative genes in both varieties, we then examined changes in the expression of major carbohydrate metabolic, glycolysis and fermentative genes (Figure 6; Table S3). Starch degradation genes including *beta-amylase* (*A_DN40578_c6_g3_i1*), *starch phosphorylase* (*SP*; *A_DN40764_c7_g1_i2*), *fructokinase* (*A_DN41293_c1_g8_i1*), and *invertase* (*A_DN40864_C7_g2_i1*) were downregulated in "A408". However, the expression of *sucrose synthase* (*SUSY*) was upregulated in both "A408" (*A_DN40966_c1_g1_i8*) and "Bali" (*B_DN52186_c2_g2_i4* and *i11*). Several genes encoded for glycolysis enzymes were strongly upregulated in "Bali", including *aldolase* (*B_DN50672_c0_g4_i1* and *i2*), *enolase* (*B_DN51208_c1_g1_i4* and *i9*), *glucose 6 phosphate (G6P) isomerase* (*B_DN50580_c2_g7_i2* and *i8*), *GAP-DH* (*B_DN51637_c1_g4_i1* and *i2*), *phosphofructokinases* (*PFKs*; *B_DN51114_c0_g1_i6* and *i9* and *B_DN51080_c4_g2_i5*), *phosphoglycerate mutase* (*PGM*; *B_DN50865_c2_g2_i5*), and *pyruvate kinases* (*PKs*; *B_DN52171_c0_g5_i1*, *i7* and *i9*).

On the other hand, the analysis of "A408" DEGs revealed only one glycolysis gene, *phospho-enol-pyruvate carboxylase kinase* (*PPCK*; *A_DN36906_c0_g1_i4*), which was induced by WS. Interestingly, *PFK* (*A_DN40730_c0_g2_i10*), encoded for one of the most important regulatory enzymes of glycolysis, was strongly downregulated under WS in "A408". Several fermentative genes were strongly upregulated in "Bali", particularly *alcohol dehydrogenases* (*ADH*; *B_DN51037_c2_g1_i1*, *i2*, *i4*, and *i5* and *B_DN50984_c1_g3_i5*), *aldehyde dehydrogenase* (*B_DN50511_c1_g7_i2*), *lactate dehydrogenase* (*LDH*; *B_DN52281_c1_g2_i1*) and *pyruvate decarboxylase* (*PDC*; *B_DN50426_c2_g1_i5* and *i10*, *B_DN50426_c2_g7_i1* and *i2*, and *B_DN52571_c0_g1_i2* and *i3*). In contrast, only two genes encoding for *ADH* (*A_DN39747_c0_g4_i3* and *A_DN40875_c0_g1_i2*) were upregulated in "A408". Our results demonstrated that genes involved in starch degradation, glycolysis, and fermentation are differentially expressed at a significantly higher level under WS in "Bali" than in "A408", suggesting that "A408" could have a slower glycolytic process and a better ability to maintain carbohydrate reserves than "Bali".

Analysis of total soluble carbohydrates in the roots of both varieties confirmed that WS resulted in a greater reduction of the total soluble carbohydrate in "Bali" than that of "A408" (Figure S1). Our results correlate with a study of wild relatives of *Arabidopsis* in the genus *Rorippa*, showing that starch degradation, glycolysis, and fermentative genes are more strongly induced in the less flooding-tolerant

R. amphibiathan than in *R. sylvestris* [36], thus suggesting that the management of carbohydrate reserves may be necessary for the survival of plants experiencing an energy crisis from low oxygen conditions.

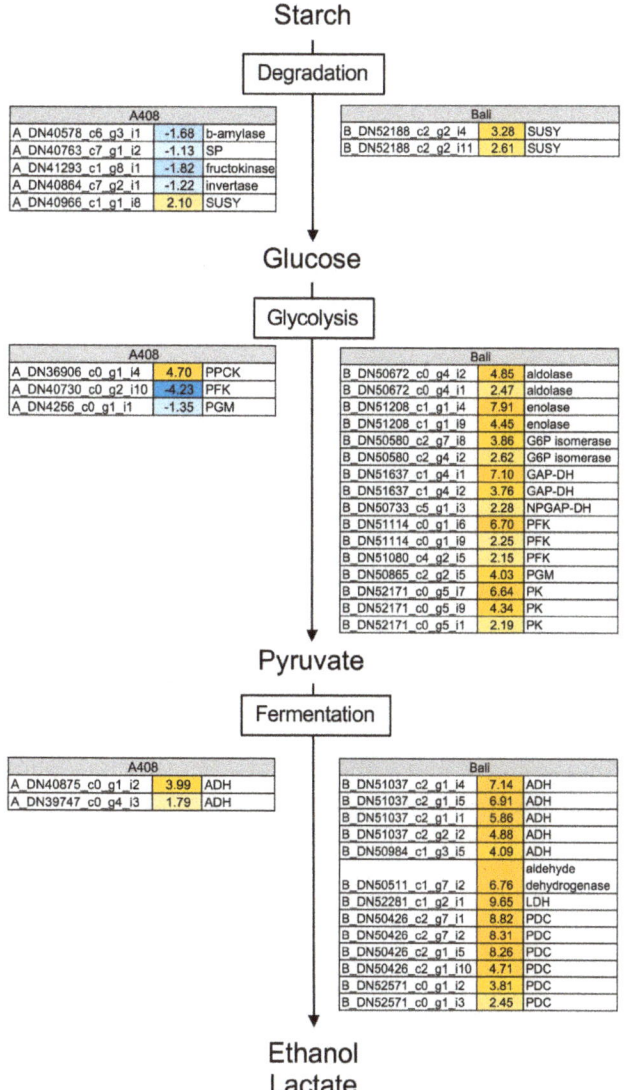

Figure 6. Waterlogging caused differential expression of major carbohydrate metabolism, glycolysis, and fermentative genes in roots of "A408" and "Bali". The number indicates log$_2$ fold changes. Blue indicates down-regulation. Yellow indicates up-regulation. Data can be found in Table S3.

2.5. Waterlogging Resulted in Stronger Induction of Ethylene Synthesis, Perception and Responsive Genes in "Bali" than in "A408"

During soil waterlogging, ethylene acts as a primary signal controlling morphological and metabolic adjustments in plant roots. Therefore, we examined the changes in gene expression of ethylene synthesis, perception, and responsive transcriptional regulator genes. In both varieties, changes in the expression of ethylene synthesis, perception, and responsive transcriptional regulatory genes were observed (Figure 7; Table S3). In "A408" one aminocyclopropane carboxylate oxidase gene (*ACC oxidase*; *A_DN37162_c1_g2_i3*), a key-enzymatic gene controlling ethylene synthesis, was upregulated and three were downregulated (*A_DN38534_c8_g1_i3*, *A_DN41071_c0_g1_i1*, and *A_DN40256_c1_g1_i1*). On the contrary, two ACC oxidase genes (*B_DN48281_c0_g1_i1* and *B_DN48281_c0_g2_i1* and *i2*) were strongly upregulated, and one was downregulated (*B_DN51722_c0_g2_i2*) in "Bali". For ethylene signaling and perception, *ERF95* (*A_DN40876_c5_g6_i1*) and *ERF106* (*A_DN34489_c0_g2_i1*) were upregulated in "A408" and *ERF2* (*B_DN48139_c0_g1_i1*) and *ERF106* (*B_DN51699_c3_g8_i1*) were upregulated in "Bali". The down-regulation of *ERF13* genes was observed in both varieties ("A408"; *A_DN45302_c0_g1_i1*, *A_DN38825_c5_g14_i1*, and *A_DN39366_c0_g1_i2* and "Bali"; *B_DN51809_c1_g5_i1*). Interestingly, the down-regulation of *ERF109* (*A_DN39318_c1_g3_i2*), a redox responsive transcription factor 1, was observed only in "A408". In *Arabidopsis*, *ERF109* is highly responsive to jasmonic acid and functions in the regulation of lateral root formation by mediating cross-talk between jasmonic acid signaling and auxin biosynthesis [37]. The expression of other ethylene-responsive transcription factor genes was generally strongly induced in "Bali", including *ERF110* (*B_DN49322_c2_g10_i2*), *DREB2C* (*B_DN44544_c1_g1_i1* and *i2*) and *Ethylene Response DNA binding factor 1* (*B_EDF1*; *DN50249_c2_g7_i1*). In "A408", *ERF subfamily B4* (*A_DN39655_c3_g1_i1*), *ERF114* (*A_DN39818_c4_g4_i1*), *RAP2.7* (*A_DN39022_c4_g2_i1*) and *RAP2.3* (*A_DN39022_c4_g2_i1*) were upregulated by WS. On the other hand, *DREB* transcription factors (*A_DN38707_c2_g6_i1*, *A_DN35768_c0_g1_i1*, and *A_DN739_c0_g1_i1*) were downregulated in "A408".

			A408
A_DN37162_c1_g2_i1	4.24	synthesis	ACC oxidase
A_DN38534_c8_g1_i3	-2.27	synthesis	ACC oxidase
A_DN41071_c0_g1_i8	-2.32	synthesis	ACC oxidase
A_DN40256_c1_g1_i1	-2.41	synthesis	ACC oxidase
A_DN40876_c5_g6_i1	2.80	perception	ERF95, ETHYLENE AND SALT INDUCIBLE 1
A_DN34489_c0_g2_i1	2.32	perception	DECREASE WAX BIOSYNTHESIS2, DEWAX2, ERF106
A_DN45302_c0_g1_i1	-3.66	perception	ERF13
A_DN38825_c5_g14_i1	-4.32	perception	ERF13
A_DN39366_c0_g1_i2	-4.45	perception	ERF13
A_DN39318_c1_g3_i2	-4.95	perception	ERF109, REDOX RESPONSIVE TRANSCRIPTION FACTOR 1
A_DN39655_c3_g1_i1	2.18	transcriptional regulator	ERF (ethylene response factor) subfamily B-4
A_DN39818_c4_g4_i1	1.64	transcriptional regulator	ERF114
A_DN39233_c3_g5_i3	1.63	transcriptional regulator	RAP2.7, TARGET OF EARLY ACTIVATION TAGGED (EAT) 1, TOE1
A_DN39022_c4_g2_i1	1.23	transcriptional regulator	RAP2.3, ERF72
A_DN38660_c8_g2_i1	-1.41	transcriptional regulator	CYTOKININ RESPONSE FACTOR 4
A_DN38707_c2_g6_i1	-2.62	transcriptional regulator	DREB1A
A_DN36824_c0_g1_i1	-4.87	transcriptional regulator	ORA47, (octadecanoid-responsive AP2/ERF-domain transcription factor 47)
A_DN40230_c1_g7_i1	-5.03	transcriptional regulator	DDF1, DWARF AND DELAYED FLOWERING 1
A_DN35768_c0_g1_i1	-5.17	transcriptional regulator	DREB subfamily A-4
A_DN26922_c0_g2_i1	-5.36	transcriptional regulator	DDF1, DWARF AND DELAYED FLOWERING 1
A_DN739_c0_g1_i1	-6.84	transcriptional regulator	DREB subfamily A-5
A_DN40040_c0_g2_i1	-8.29	transcriptional regulator	ERF17

			Bali
B_DN48281_c0_g2_i2	7.69	synthesis	ACC oxidase
B_DN48281_c0_g2_i1	5.07	synthesis	ACC oxidase
B_DN48281_c0_g1_i1	3.21	synthesis	ACC oxidase
B_DN51722_c0_g2_i2	-5.11	synthesis	ACC oxidase
B_DN47749_c0_g2_i2	-7.48	synthesis	2-oxoglutarate (2OG) and Fe(II)-dependent oxygenase superfamily protein
B_DN45914_c0_g1_i1	3.99	perception	PPPDE thiol peptidase family protein
B_DN51699_c3_g8_i1	3.12	perception	DECREASE WAX BIOSYNTHESIS2, DEWAX2, ERF106
B_DN48139_c0_g1_i1	2.82	perception	ERF2
B_DN45520_c0_g1_i1	-2.06	perception	ERF1
B_DN50829_c0_g2_i8	-3.16	perception	PPPDE thiol peptidase family protein
B_DN51809_c1_g5_i1	-3.28	perception	ERF13
B_DN51798_c0_g7_i3	-4.96	perception	PPPDE thiol peptidase family protein
B_DN2643_c0_g1_i1	-7.03	perception	ERF (ethylene response factor) subfamily B-1
B_DN49322_c2_g10_i2	8.33	transcriptional regulator	ERF110
B_DN44544_c0_g1_i2	5.92	transcriptional regulator	DREB2C
B_DN44544_c0_g1_i1	2.75	transcriptional regulator	DREB2C
B_DN50249_c2_g7_i1	1.88	transcriptional regulator	ATTEM1, EDF1, ETHYLENE RESPONSE DNA BINDING FACTOR 1, TEM1, TEMPRANILLO 1
B_DN51305_c2_g1_i1	-5.65	transcriptional regulator	RAP2.11

Figure 7. Differential expression pattern of ethylene synthesis, perception and transcriptional regulator genes in roots of "A408" and "Bali" subjected to WS. The number indicates $\log_2$ fold changes. Blue indicates down-regulation. Yellow indicates up-regulation. Data can be found in Table S3.

The stronger up-regulation of *ACC oxidase* could result in higher ethylene production in "Bali" than in "A408". Ethylene has an important role during lateral root initiation as treatment of ethylene reduces lateral root initiation in *Arabidopsis* seedlings [38]. Moreover, *Arabidopsis* mutants with enhanced ethylene synthesis or perception decreased lateral root formation, while ethylene-insensitive mutants increased the number of lateral roots [38]. Furthermore, Muday et al. [39] discussed the antagonistic roles of auxin and ethylene in controlling lateral root formation, in which the control of lateral root development by ethylene involves changes in auxin transport and accumulation patterns [40].

2.6. Auxin Metabolism and Auxin-Regulated Transcription Factor Genes were Predominantly Induced in the Roots of Waterlogging-Tolerant Zombi Pea

Auxin participates in root growth and the regulation of lateral root development. Our phenotypic data demonstrated that WS resulted in adaptive changes of zombi pea root phenotypes (Figure 2). GO enrichment analysis also suggests the down-regulation of IAA carboxyl methyltransferase activity in WS "Bali" roots (Figure 4). Recently, IAA methylation was proposed to function in maintaining auxin homeostasis by regulating the polar auxin transportation [41]. Moreover, the *AUX/IAA* family is overrepresented in the upregulated DEGs of "A408" based on Fisher's exact test for over-representation analysis (Figure 5; Table S4). Therefore, we examined the changes in the expression of auxin metabolism and auxin-responsive transcription factor genes. Under WS, genes involved in auxin metabolism and auxin-responsive transcription factor genes were differentially regulated in both varieties (Figure 8; Table S3). However, four *Small Auxin Upregulated RNAs* (*SAURs*) were induced in "A408" (*A_DN2515_c0_g1_i1*, *A_DN38724_c0_g4_i1*, *A_DN40333_c0_g1_i1*, and *A_DN40413_c3_g11_i1*). In contrast, only one *SAUR* (*B_DN52605_c0_g2_i1*) was induced in "Bali". Evidently, *SAURs* can regulate auxin-induced acid growth as defined by the loosening of cell walls at low pH which promotes cell wall extensibility and rapid cell elongation [42]. Our results demonstrate that WS in "A408" can upregulate a SAUR gene, *A_DN40413_c3_g11_i1*. The best BLAST hit of *A_DN40413_c3_g11_i1* protein is the *Arabidopsis* SAUR51 (AT1G75580; Table S3). Previous studies in *Arabidopsis* demonstrated that *SAUR51* is an auxin-inducible gene [43] which is highly expressed in root primordia [44], which suggests it may function in lateral root growth under WS.

		A408	
A_DN38724_c0_g4_i1	4.45	auxin metabolism	SAUR14, SMALL AUXIN UPREGULATED RNA 14
A_DN2515_c0_g1_i1	4.14	auxin metabolism	SAUR14, SMALL AUXIN UPREGULATED RNA 14
A_DN39053_c2_g11_i1	3.67	auxin metabolism	IAA-amido synthase, GH3.1
A_DN40413_c3_g11_i1	2.88	auxin metabolism	SAUR51, SMALL AUXIN UPREGULATED RNA 51
A_DN40333_c0_g1_i1	1.26	auxin metabolism	SAUR55, SMALL AUXIN UPREGULATED RNA 55
A_DN39619_c1_g1_i1	1.02	auxin metabolism	O-fucosyltransferase family protein
A_DN40136_c1_g2_i1	-1.47	auxin metabolism	Auxin-responsive family protein
A_DN40630_c1_g3_i2	-1.50	auxin metabolism	Auxin-responsive family protein
A_DN38365_c3_g8_i4	-2.05	auxin metabolism	LCV3, LIKE COV 3
A_DN39227_c1_g4_i2	6.51	transcriptional regulator	IAA14, INDOLE-3-ACETIC ACID INDUCIBLE 14, SLR, SOLITARY ROOT
A_DN38749_c3_g1_i3	1.96	transcriptional regulator	SHY2/IAA3
A_DN41048_c2_g4_i1	1.65	transcriptional regulator	ATAUX2-11, AUXIN INDUCIBLE 2-11, IAA4, INDOLE-3-ACETIC ACID INDUCIBLE 4
A_DN39227_c1_g4_i1	1.49	transcriptional regulator	IAA14, INDOLE-3-ACETIC ACID INDUCIBLE 14, SLR, SOLITARY ROOT
A_DN41048_c2_g1_i1	1.30	transcriptional regulator	ATAUX2-11, AUXIN INDUCIBLE 2-11, IAA4, INDOLE-3-ACETIC ACID INDUCIBLE 4

		Bali	
B_DN52605_c0_g2_i1	7.88	auxin metabolism	SAUR14, SMALL AUXIN UPREGULATED RNA 14
B_DN50148_c6_g3_i4	-2.97	auxin metabolism	NAD(P)-linked oxidoreductase superfamily protein
B_DN52035_c2_g3_i26	-4.32	auxin metabolism	IAA-amido synthase, DFL1, DWARF IN LIGHT 1, GH3.6, GRETCHEN HAGEN3.6
B_DN49797_c4_g3_i1	-2.96	auxin metabolism	CYP711A1, CYTOCHROME P450, FAMILY 711, SUBFAMILY A, POLYPEPTIDE 1
B_DN47104_c0_g1_i1	-2.53	auxin metabolism	ATIAMT1, IAA CARBOXYLMETHYLTRANSFERASE 1, IAMT
B_DN52121_c6_g8_i1	3.94	transcriptional regulator	ATAUX2-11, AUXIN INDUCIBLE 2-11, IAA4, INDOLE-3-ACETIC ACID INDUCIBLE 4

Figure 8. Differential expression pattern of auxin metabolism and transcriptional regulator genes in roots of "A408" and "Bali" subjected to WS. The number indicates $\log_2$ fold changes. Blue indicates down-regulation. Yellow indicates up-regulation. Data can be found in Table S3.

Additionally, we observed the up-regulation of two key regulators in auxin-regulated lateral root development, *SHORT HYPOCOTYL 2/SUPPRESSOR OF HY 2 (SHY2)/Indole-3-acetic acid-inducible (IAA) 3* (A_DN38749_c3_g4_i1) and *IAA14* (A_DN39227_c1_g4_i1 and i2) only in the DEGs of "A408". SHY2/IAA3 and IAA14 are auxin-inducible transcriptional repressors that regulate auxin-mediated gene expression by controlling the activity of auxin response factors (ARFs) by protein-protein interaction [45–47]. Goh et al. [48] proposed that multiple *Aux/IAA–ARF* modules cooperatively regulate the developmental steps during lateral root formation. Therefore, we speculated that *SHY2/IAA3* and *IAA14* could specifically regulate zombi pea lateral root formation under WS. An in-depth analysis of the WS-induced *SAURs* and *Aux/IAAs* is required to further provide candidate genes for improving waterlogging tolerance in *Vigna* crops.

2.7. Differential Expression of Abscisic Acid and Gibberellic Acid Metabolic Genes

Abscisic acid (ABA) and gibberellic acid (GA) play antagonistic roles to control plant development and response to environmental stresses. GO enrichment analysis also suggests the down-regulation of farnesyltranstransferase activity and gibberellin 3-beta-dioxygenase activity in WS "Bali" roots (Figure 4). Additionally, ORA identified overrepresentation of GA metabolism genes in the downregulated DEGs of WS "Bali" roots (Figure 5). The biosynthesis of ABA and GA both derives from the isoprenoid pathway. Cutler et al. [49] demonstrated that *farnesyl transferase* is a key regulator of ABA signal transduction in *Arabidopsis*. Interestingly, the down-regulation of *farnesyl transferase* increases the ABA response and drought tolerance in *Brassica napa* [50]. In this study, the down-regulation of *genenylgeranyl pyrophosphate synthase 1* (*GGPS1*; B_DN993_c0_g1_i1) was observed in WS "Bali" roots (Table S3). *Arabidopsis* GGPS1 (encoded by *At4g36810*) has farnesyl transferase activity and functions as a key enzyme in the chloroplast isoprenoid biosynthetic pathway. GGPS1 catalyzes the formation of geranylgeranyl diphosphate, a precursor molecule of carotenoids, ABA, and GA [51]. Moreover, the down-regulation of *GA 3-oxidase 1* (*GA3OX1*: B_DN46875_c0_g1_i1) was also found in WS "Bali" roots (Table S3). *Arabidopsis GA3OX1* is involved in the production of bioactive GA, and plays an essential role in the regulation of root growth [52]. Altogether, these results suggest the modulation of ABA and GA level could play a role in the regulation of waterlogging tolerance in zombi pea.

2.8. Differential Expression of Transport Genes

In general, most of the transporter gene families were downregulated in response to WS in both varieties (Figure 9A,B), except for a family of major intrinsic protein (aquaporin) genes which were largely induced under WS in "A408" (Figure 9A,B). The aquaporin has an important role in the regulation of plant water uptake, water loss, and hydraulic conductivity [53]. Among these aquaporin genes, three *plasma membrane intrinsic proteins* (PIPs) (A_DN39755_c1_g1_i7, A_DN40480_c0_g1_i6, and A_DN40480_c0_g3_i1) were specifically induced in "A408" (Figure 9C; Table S3). The best BLAST hit of A_DN39755_c1_g1_i7 protein is the *Arabidopsis PIP2;7* (AT4G35100; Table S3). Functional analysis of the *Arabidopsis PIP2;7* revealed that it is highly expressed in root elongating cells, and is most likely involved in cell elongation processes where the regulation of water movement is crucial [54]. Taken together, these results suggest the upregulation of aquaporin genes may contribute to waterlogging tolerance in the zombi pea.

2.9. Differential Expression of Plant Cell Wall-Related Genes

Since the aerenchyma formation was observed in WS roots and hypocotyl of "A408" (Figure 3A,B), we sought to determine the change in expression of cell wall-related genes upon WS. Our results demonstrated that cell wall-related genes were overrepresented in the upregulated DEGs of "A408" (Figure 5; Table S4). Genes involved in cell wall modification were overrepresented in the upregulated DEGs of both "A408" and "Bali" (Table S4). However, the group of pectin methylesterase genes, including both *pectin methylesterases* and *pectin methylesterase inhibitors*, were specifically overrepresented in the upregulated DEGs of "A408" (Figure 10A,B; Table S4). Pectin is a structurally compact polysaccharide that is a constituent of plant's primary cell wall. Pectin plays a key role in plant growth, cell expansion, and response to stress [55]. Pectin methylesterases and pectin methylesterase inhibitors are enzymes involved in shoot apical meristem development and root tip elongation through plant hormone pathways including auxin [56,57]. Our results suggest that the modification of the plant cell walls by pectin methylesterases and pectin methylesterase inhibitors could have a role in the waterlogging tolerance of zombi pea. In support of this, *Glyma.03.g029400*, a soybean root-specific *pectin methylesterase inhibitor*, has been proposed as the likely underlying gene of a major QTL for waterlogging tolerance, *qWT_Gm03* [17].

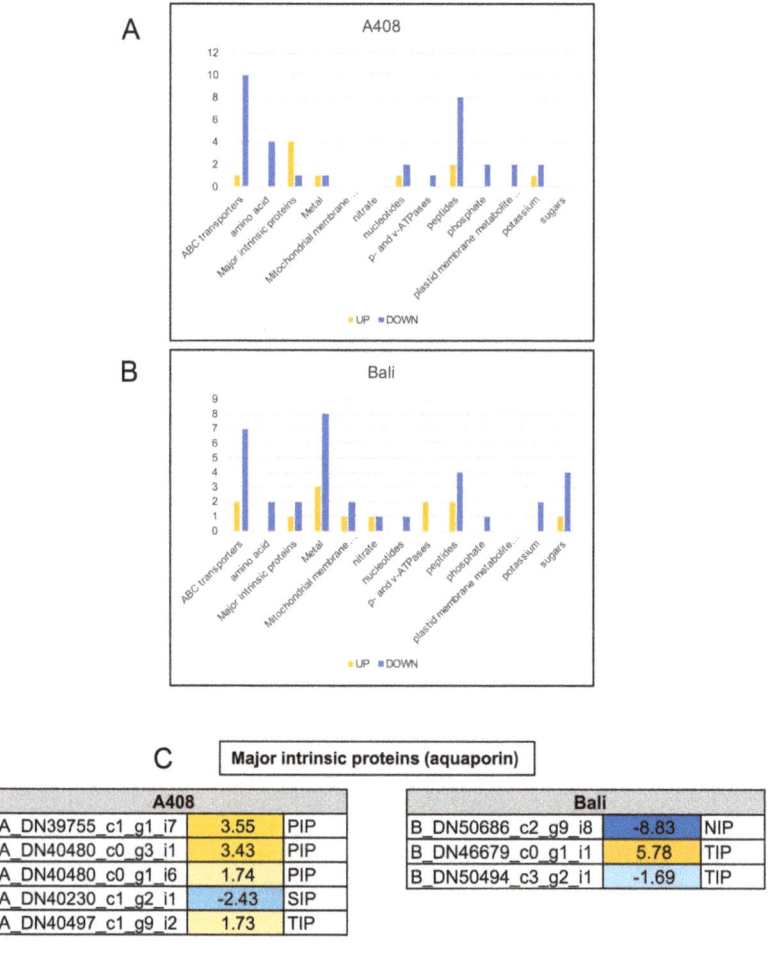

Figure 9. Differential expression of transport genes in roots of "A408" and "Bali" subjected to WS. Graphical representation of WS-regulated transport genes based on their assigned protein families. "Up" and "Down" represent up-regulation and down-regulation in this analysis. (**A**) "A408". (**B**) "Bali". (**C**) Expression patterns of major intrinsic protein (aquaporin) genes in roots of "A408" and "Bali" under WS. The number indicates log$_2$ fold changes. Blue indicates down-regulation. Yellow indicates up-regulation. Data can be found in Table S3.

Cell wall-associated peroxidases are enzymes that use hydrogen peroxide and/or superoxide anions as substrates to catalyze a production of hydroxyl radicals. The production of hydroxyl radicals can cause an increase in cell wall loosening during auxin-mediated cell wall extension. Here, we observed the over-representation of peroxidase genes in the upregulated DEGs of "A408" (Figure 5; Table S4). In the DEGs of "A408", nine out of 11 peroxidase DEGs were upregulated under WS (Figure 10C). In contrast, only five out of 16 peroxidase DEGs were upregulated in the DEGs of "Bali" (Figure 10C). Interestingly, the expression of $A\_DN39902\_c2\_g1\_i1$ and $A\_DN40709\_c4\_g1\_i2$ was upregulated only in the DEGs of "A408" (Figure 10C). The best blast hit of these two transcripts is the *Arabidopsis* cell wall loosening peroxidase 53 (Prx53: AT5G06720; Table S3) [58]. The results suggest that these genes might play some role in waterlogging response.

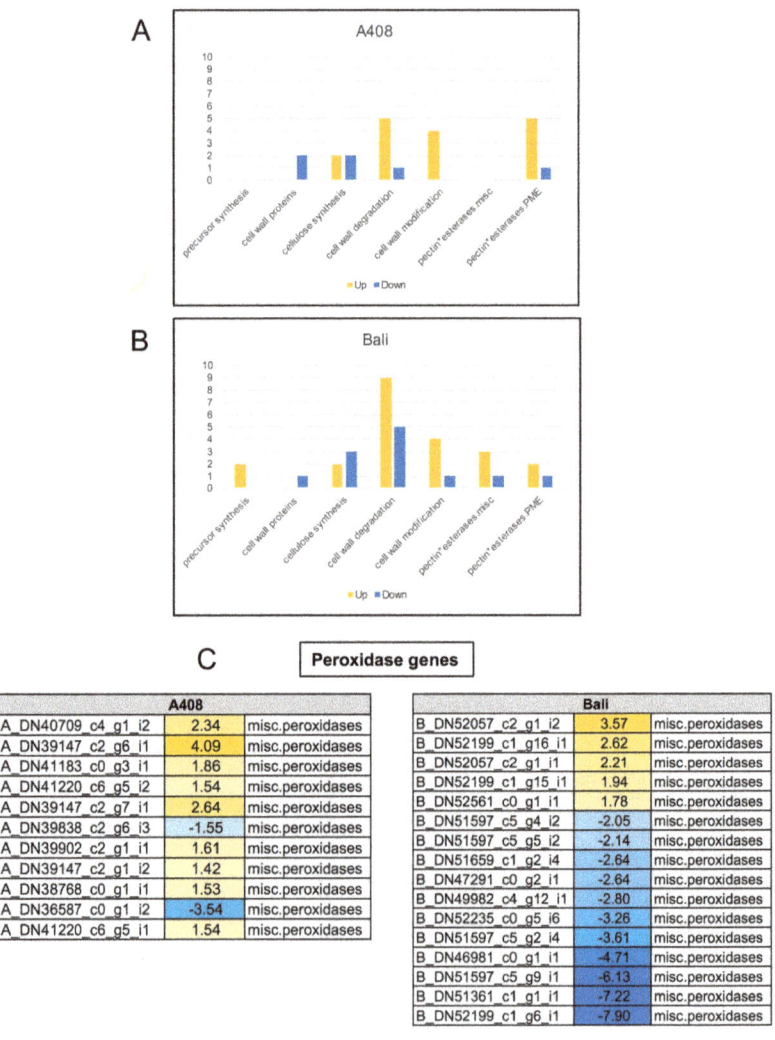

Figure 10. Differential expression of cell wall-related and peroxidase genes in roots of "A408" and "Bali" subjected to WS. Graphical representation of WS-regulated cell wall-related genes based on their assigned protein families. "Up" and "Down" represent up-regulation and down-regulation in this analysis. (**A**) "A408". (**B**) "Bali". (**C**) Expression patterns of peroxidase genes in roots of "A408" and "Bali" under WS. The number indicates $\log_2$ fold changes. Blue indicates down-regulation. Yellow indicates up-regulation. Data can be found in Table S3.

2.10. Validation of Transcriptome Data by Quantitative Real-Time PCR

To validate the transcriptome results, for each variety we selected six DEGs and one non-DEG based on their function and expression level for quantitative real-time PCR analysis (qRT-PCR). For "A408", of the six DEGs, five genes were upregulated including *pectin lyase-like superfamily protein (A_DN38641_c0_g1_i3), SUSY (A_DN40966_c1_g1_i8), ADH (A_DN39747_c0_g4_i3), aquaporin tonoplast intrinsic protein (TIP; A_DN40497_c1_g9_i2),* and *IAA14 (A_DN39227_c1_g4_i1)* and one DEGs, *WRKY transcription factor (A_DN40719_c0_g11_i2)* was downregulated (Figure S2A). For "Bali", of the six DEGs, four genes were upregulated including *ADH (B_DN50984_c1_g3_i5), GAP-DH (B_DN51637_c1_g4_i2), glucose-6-phosphate isomerase (B_DN50580_c2_g4_i2), aldolase (B_DN50672_c0_g4_i1)* and two DEGs, *TIP (B_DN50494_c3_g2_i1)* and *auxin-induced protein PCNT115 (B_DN50148_c6_g3_i4)* were downregulated (Figure S2B). The expression of a non-DEG, *ATP synthase subunit beta* ("A408"; *A_DN39747_c0_g13_i1* and "Bali"; *B_DN50009_c1_g1_i3*), was used as a reference for the relative gene expression calculation. Our qRT-PCR results demonstrate the reliability of the RNA-seq data.

3. Material and Methods

3.1. Plant Material and Stress Treatment

Vigna vexillata seeds (JP235863 ("Bali") and AusTRCF320047 ("A408") varieties) were germinated in soil and grown outdoors between April and June of 2016 and 2017 at Kasetsart University, Bang Khen campus. Fifteen-day-old, five-leaf-stage plants were used in the WS treatment. In brief, plant pots were placed in plastic containers filled with tap water. The level of water was set at 3 cm above the soil. Waterlogging stress began at midday and continued for 24 h. For the control, non-treated plants were placed in a container with no water. For each sample, the root tissue of five plants was harvested at the end of the treatment; it was immediately placed in liquid nitrogen, ground into a fine powder, and kept at −80 °C. For long term WS, plants were subjected to WS for up to 10 days.

3.2. Analysis of Leaf Chlorophyll Content

Adhering to the method described by Juntawong et al. [2], chlorophyll content was measured using the atLEAF+ chlorophyll meter (FT Green LLC, Wilmington, DE). The youngest fully-expanded leaves were measured three times at 10.00 am and the averages were used in subsequent analysis. Twelve plants were analyzed for each time point. The total chlorophyll content of the leaves was obtained by converting the atLEAF+ values in SPAD units and the total chlorophyll contents using an online web tool: http://www.atleaf.com/SPAD.aspx.

3.3. Root Anatomical and Morphological Analysis

For analyses of root morphology, underground roots were collected and photographed after seven days of WS. Roots of NS plants grown side by side were used as controls. For the anatomical study, taproots and hypocotyls were immediately fixed in formaldehyde-acetic acid-alcohol (FAA) solution. Permanent slides for microscopic observation were prepared by standard microtechnique procedures [59]. The embedded samples were sectioned at 10–15 micrometer thickness using a rotary microtome (Leica RM2165; Leica Biosystems, Germany) and stained with Safranin and Fast Green. The samples were observed under a bright-field microscope (Axioskop 2 Plus; Zeiss, Germany) equipped with a digital camera (AxioCam MRc; Zeiss, Germany).

3.4. Analysis of Total Soluble Carbohydrate Content

One hundred milligrams of frozen root tissue was used to quantify the total soluble carbohydrate content using a method described by Juntawong et al. [2]. In brief, soluble carbohydrates were extracted, hydrolyzed by adding 5 mL of 2.5 N HCl, and incubated in a boiling water bath for 3 h. The addition of 0.75 g of Na_2CO_3 was performed to neutralize the extract. The anthrone method was used to determine total carbohydrate content relative to a standard series of glucose. In brief, the extract (300 µL) and distilled water (700 µL) were mixed with 4 mL of 0.14% (w/v) anthrone solution in 95% H_2SO_4; it was then incubated in a boiling water bath for 8 min, and rapidly cooled on ice. The absorbance was quantified at 630 nm.

3.5. RNA Extraction, Library Preparation, and Sequencing

Total RNA was extracted with TRIzol reagent (Invitrogen), according to the manufacturer's protocol. Total RNA samples were subjected to DNase treatment and RNA cleanup using an RNA-mini kit (Qiagen). Two replicates of total RNA samples were used for transcriptome analysis according to the ENCODE recommended RNA-seq standards (https://genome.ucsc.edu/ENCODE/protocols/dataStandards/ENCODE_RNAseq_Standards_V1.0.pdf). The integrity of the RNA samples (RIN) was evaluated on an RNA 6000 Nano LapChiprun on Agilent2100 Bioanalyzer (Agilent Technologies, Germany). Samples with a RIN > 7 were used in RNA-seq library preparation. One µg of total RNAs were used to generate a sequencing library using a NEBNext® Ultra™ RNA Library Prep Kit for Illumina® following the manufacturer's instructions. The mRNA fragmentation and priming were performed using NEBNext First Strand Synthesis Reaction Buffer and NEBNext Random Primers. First-strand cDNA was synthesized using ProtoScript II Reverse Transcriptase and the second-strand cDNA was synthesized using Second Strand Synthesis Enzyme Mix. The purified (via AxyPrep Mag PCR Clean-up (Axygen)) double-stranded cDNA was then treated with End Prep Enzyme Mix to repair both ends and added a dA-tailing in one reaction, followed by a T-A ligation to add adaptors to both ends. Size selection of Adaptor-ligated DNA was then performed using AxyPrep Mag PCR Clean-up (Axygen) and fragments of ~360 bp (with the approximate insert size of 300 bp) were recovered. Each sample was then amplified by PCR for 11 cycles using P5 and P7 primers, with both primers carrying sequences that could anneal with flow cell to perform bridge PCR and P7 primer carrying a six-base index allowing for multiplexing. The PCR products were cleaned using AxyPrep Mag PCR Clean-up (Axygen), validated using an Agilent 2100 Bioanalyzer (Agilent Technologies, Palo Alto, CA, USA), and quantified using Qubit 2.0 Fluorometer (Invitrogen, Carlsbad, CA, USA). Then libraries with different indices were multiplexed and loaded on an Illumina HiSeq 4000 instrument according to the manufacturer's instructions (Illumina, San Diego, CA, USA). Sequencing was carried out using a 2 × 150 bp paired-end (PE) configuration; image analysis and base calling were conducted by the HiSeq Control Software (HCS) + RTA 2.7 (Illumina) on the HiSeq 4000 instrument. The raw read files were deposited in the NCBI SRA database under the accession numbers SRR9214917-SRR9214924. Quality control filtering and 3' end trimming were analyzed using the FASTX-toolkit (http://hannonlab.cshl.edu/fastx_toolkit/index.html) and Trimmomatic software [60], respectively.

3.6. De Novo Assembly and Annotation

The transcriptome was assembled using the Trinity software (https://github.com/trinityrnaseq/trinityrnaseq) [61]. The assembly was performed using a k-mer value of 25 with default parameters. The de novo transcriptome assembled files can be found in Files S1 and S2. The protein sequences derived from the assembled transcriptomes were further annotated using BLASTP to plant UniprotPK database with an E value threshold of 1e−10 using AgBase (http://agbase.arizona.edu) and the Mercator annotation pipeline with a blast cut-off score of 80 (https://plabipd.de/portal/mercator-sequence-annotation). The annotation information can be found in Table S2.

3.7. Differential Gene Expression Analysis

Differential gene expression analysis was performed according to Sirikhachornkit et al. [62]. The FASTQ files were aligned to the reference transcriptome using Bowtie2 software (http://bowtie-bio.sourceforge.net/bowtie2/index.shtml). A binary format of sequence alignment files (BAM) was generated and used to create read count tables using the HTseq python library (https://htseq.readthedocs.

io/). Differentially-expressed genes were calculated using the edgeR program (https://bioconductor.org/packages/release/bioc/html/edgeR.html) with an FDR cutoff of < 0.05.

Gene ontology enrichment analysis was performed in the R environment using the GOHyperGAll function [63]. Gene annotation files were generated using the AgBase webtool. Significant GO terms were filtered by an adjusted *p*-value of < 0.05.

For PAGEMAN analysis, the mapping file was generated from the protein sequences derived from the assembled transcriptomes using the Mercator pipeline. Over-representation analysis (ORA) was performed using the PAGEMAN [64] program with Fisher's test and a cutoff value of two.

Homolog identification was performed using translated amino acid sequences (>100 amino acids) derived from the transcriptomes of "A408" and "Bali" and *A. thaliana* protein sequences (TAIR10) by OrthoVenn2 [65]. The homolog clusters and expression can be found in Table S3.

3.8. Quantitative Real-Time PCR

Three replicates of total RNA samples were used. Total RNAs were treated with DNase I (NEB, USA) to eliminate contaminated genomic DNA. One microgram of total RNAs were used to construct cDNA using MMuLv reverse transcriptase (Biotechrabbit, Germany) in a final volume of 20 µL. The cDNA was diluted five times. Quantitative-realtime PCR (qPCR) reaction was performed according to Sirikhachornkit et al. [62]. Further, qPCR was performed using QPCR Green Master Mix (Biotechrabbit, Germany) on a MasterCycler RealPlex ^{4} (Eppendorf, Germany). For each sample, the PCR reaction was performed in triplicate. Each reaction contained 1 µL of diluted cDNA, 0.5 µM of each primer, and 10 µL of QPCR Green Master Mix, giving a final volume of 20 µL. The PCR cycle was 95 °C for 2 min, followed by 45 cycles of 95 °C for 15 s and 60 °C for 30 s. Amplification specificity was validated by melt-curve analysis at the end of each PCR experiment. Relative gene expression was calculated using the $2^{-\Delta\Delta CT}$ method. The genes and primers used are shown in Table S5.

4. Conclusions

In this research, we aimed to discover the molecular mechanisms controlling waterlogging tolerance by constructing de novo transcriptomes and comparing the transcriptomes of two zombi pea varieties with contrasting waterlogging tolerance. Our results demonstrated that root plasticity could be an important determinant factor controlling waterlogging tolerance in zombi pea. Moreover, differential expressions of multiple genes encoding for energy production pathways, auxin-regulated lateral root initiation and formation, hormones, cell wall modification, membrane transporter, and peroxidase could contribute to waterlogging tolerance in zombi pea. Functional characterization of the WS-induced candidate genes is required to further identify candidate genes controlling waterlogging-tolerant traits. Additionally, recent studies demonstrated that differentially-regulated genes controlling for the traits of interest could be accurately identified using comparative transcriptome RNA-seq analysis of near-isogenic lines (NILs) [66,67]. Clearly, this method could help to narrow down the list of candidate genes responsible for waterlogging tolerance in zombi pea by removing genetic background effects. We expect that the basic knowledge obtained from this study will be used to help design further experiments focused on improving our understanding of the morphological and physiological responses to waterlogging that are important for molecular breeding of waterlogging-tolerant *Vigna* crops in the future.

Supplementary Materials: The following are available online at http://www.mdpi.com/2223-7747/8/8/264/s1, File S1: "A408" de novo assembled transcriptome data. File S2: "Bali" de novo assembled transcriptome data. Figure S1: Total root carbohydrate content. (A) "A408". (B) "Bali". Data represent mean ± SE (n = 3). * p < 0.05, ** p < 0.01 (*t*-test). Figure S2: Quantitative real-time PCR validation of transcriptome data for selected genes. (A) "A408". (B) "Bali". Relative expression was normalized to the abundance of *ATP synthase subunit beta*. Data represent mean ± SE (n = 3). * p < 0.05, ** p < 0.01 (*t*-test). Table S1: Transcriptome statistics. Table S2: Transcriptome annotation. Table S3: Differentially-expressed transcripts. Table S4: Comparative transcriptome analysis. Table S5: List of genes and primers used for qRT-PCR.

Author Contributions: Conceptualization, P.J. and P.S.; Methodology, P.B., P.J. and O.K.; Validation, P.B. and P.J.; Formal Analysis, P.J.; Investigation, P.B., P.J. and O.K.; Resources, P.S.; Writing—Original Draft Preparation, P.J.; Writing—Review & Editing, P.J.; Supervision, P.J.; Funding Acquisition, P.J.

Funding: This study was fsunded by the Faculty of Science, Kasetsart University, Kasetsart University Research and Development Institute, and the Thailand Research Fund (RSA6280013).

Acknowledgments: The authors would like to thank Pattralak Songserm and Ratchaneeporn Pimjan for technical assistant.

Conflicts of Interest: The authors declare no conflicts of interest.

References

1. Sasidharan, R.; Bailey-Serres, J.; Ashikari, M.; Atwell, B.J.; Colmer, T.D.; Fagerstedt, K.; Fukao, T.; Geigenberger, P.; Hebelstrup, K.H.; Hill, R.D.; et al. Community recommendations on terminology and procedures used in flooding and low oxygen stress research. *New Phytol.* **2017**, *214*, 1403–1407. [CrossRef] [PubMed]
2. Juntawong, P.; Sirikhachornkit, A.; Pimjan, R.; Sonthirod, C.; Sangsrakru, D.; Yoocha, T.; Tangphatsornruang, S.; Srinives, P. Elucidation of the molecular responses to waterlogging in Jatropha roots by transcriptome profiling. *Front. Plant Sci.* **2014**, *5*, 658. [CrossRef] [PubMed]
3. Bailey-Serres, J.; Fukao, T.; Gibbs, D.J.; Holdsworth, M.J.; Lee, S.C.; Licausi, F.; Perata, P.; Voesenek, L.A.; van Dongen, J.T. Making sense of low oxygen sensing. *Trends Plant Sci.* **2012**, *17*, 129–138. [CrossRef] [PubMed]
4. Fukao, T.; Xiong, L. Genetic mechanisms conferring adaptation to submergence and drought in rice: Simple or complex? *Curr. Opin. Plant Biol.* **2013**, *16*, 196–204. [CrossRef] [PubMed]
5. Voesenek, L.A.; Bailey-Serres, J. Flood adaptive traits and processes. An overview. *New Phytol.* **2015**, *206*, 57–73. [CrossRef]
6. Xu, K.; Xu, X.; Fukao, T.; Canlas, P.; Maghirang-Rodriguez, R.; Heuer, S.; Ismail, A.M.; Bailey-Serres, J.; Ronald, P.C.; Mackill, D.J. Sub1A is an ethylene-response-factor-like gene that confers submergence tolerance to rice. *Nature* **2006**, *442*, 705–708. [CrossRef]
7. Hattori, Y.; Nagai, K.; Furukawa, S.; Song, X.J.; Kawano, R.; Sakakibara, H.; Wu, J.; Matsumoto, T.; Yoshimura, A.; Kitano, H.; et al. The ethylene response factors SNORKEL1 and SNORKEL2 allow rice to adapt to deep water. *Nature* **2009**, *460*, 1026–1030. [CrossRef]
8. Rivera-Contreras, I.K.; Zamora-Hernandez, T.; Huerta-Heredia, A.A.; Capataz-Tafur, J.; Barrera-Figueroa, B.E.; Juntawong, P.; Pena-Castro, J.M. Transcriptomic analysis of submergence-tolerant and sensitive Brachypodium distachyon ecotypes reveals oxidative stress as a major tolerance factor. *Sci. Rep.* **2016**, *6*, 27686. [CrossRef]
9. Valliyodan, B.; Van Toai, T.T.; Alves, J.D.; Goulart, P.D.P.; Lee, J.D.; Fritschi, F.B.; Rahman, M.A.; Islam, R.; Shannon, J.G.; Nguyen, H.T. Expression of Root-Related Transcription Factors Associated with Flooding Tolerance of Soybean (Glycine max). *Int. J. Mol. Sci.* **2014**, *15*, 17622–17643. [CrossRef]
10. Jitsuyama, Y. Morphological root responses of soybean to rhizosphere hypoxia reflect waterlogging tolerance. *Can. J. Plant Sci.* **2015**, *95*, 999–1005. [CrossRef]
11. Kim, Y.H.; Hwang, S.J.; Waqas, M.; Khan, A.L.; Lee, J.H.; Lee, J.D.; Nguyen, H.T.; Lee, I.J. Comparative analysis of endogenous hormones level in two soybean (*Glycine max* L.) lines differing in waterlogging tolerance. *Front. Plant Sci.* **2015**, *6*, 714. [CrossRef]
12. Sakazono, S.; Nagata, T.; Matsuo, R.; Kajihara, S.; Watanabe, M.; Ishimoto, M.; Shimamura, S.; Harada, K.; Takahashi, R.; Mochizuki, T. Variation in Root Development Response to flooding among 92 Soybean Lines during Early Growth Stages. *Plant Prod. Sci.* **2014**, *17*, 228–236. [CrossRef]
13. Cornelious, B.; Chen, P.; Hou, A.; Shi, A.; Shannon, J.G. Yield Potential and Waterlogging Tolerance of Selected Near-Isogenic Lines and Recombinant Inbred Lines from Two Southern Soybean Populations. *J. Crop Improv.* **2006**, *16*, 97–111. [CrossRef]
14. Nguyen, V.T.; Vuong, T.D.; VanToai, T.; Lee, J.D.; Wu, X.; Mian, M.A.R.; Dorrance, A.E.; Shannon, J.G.; Nguyen, H.T. Mapping of Quantitative Trait Loci Associated with Resistance to Phytophthora sojae and Flooding Tolerance in Soybean. *Crop Sci.* **2012**, *52*, 2481–2493. [CrossRef]
15. Rhine, M.D.; Stevens, G.; Shannon, G.; Wrather, A.; Sleper, D. Yield and nutritional responses to waterlogging of soybean cultivars. *Irrig. Sci.* **2010**, *28*, 135–142. [CrossRef]

16. VanToai, T.T.; Beuerlein, A.F.; Schmitthenner, S.K.; St. Martin, S.K. Genetic Variability for Flooding Tolerance in Soybeans. *Crop Sci.* **1994**, *34*, 1112–1115. [CrossRef]
17. Ye, H.; Song, L.; Chen, H.T.; Valliyodan, B.; Cheng, P.; Ali, L.; Vuong, T.; Wu, C.J.; Orlowski, J.; Buckley, B.; et al. A major natural genetic variation associated with root system architecture and plasticity improves waterlogging tolerance and yield in soybean. *Plant Cell Environ.* **2018**, *41*, 2169–2182. [CrossRef]
18. Gibberd, M.R.; Gray, J.D.; Cocks, P.S.; Colmer, T.D. Waterlogging tolerance among a diverse range of Trifolium accessions is related to root porosity, lateral root formation and 'aerotropic rooting'. *Ann. Bot.* **2001**, *88*, 579–589. [CrossRef]
19. Malik, A.I.; Ailewe, T.I.; Erskine, W. Tolerance of three grain legume species to transient waterlogging. *AoB Plants* **2015**, *7*, 40. [CrossRef]
20. Lewis, G.P. *Legumes of the World*; Royal Botanic Gardens Kew: London, UK, 2005.
21. Dachapak, S.; Somta, P.; Poonchaivilaisak, S.; Yimram, T.; Srinives, P. Genetic diversity and structure of the zombi pea (*Vigna vexillata* (L.) A. Rich) gene pool based on SSR marker analysis. *Genetica* **2017**, *145*, 189–200. [CrossRef]
22. Tomooka, N.; Yoon, M.S.; Doi, K.; Kaga, A.; Vaughan, D. AFLP analysis of diploid species in the genus Vigna subgenus Ceratotropis. *Genet. Resour. Crop Evol.* **2002**, *49*, 521–530. [CrossRef]
23. Tomooka, N.; Naito, K.; Kaga, A.; Sakai, H.; Isemura, T.; Ogiso-Tanaka, E.; Iseki, K.; Takahashi, Y. Evolution, domestication and neo-domestication of the genus Vigna. *Plant Genet. Resour.* **2014**, *12*, 168–171. [CrossRef]
24. Kumar, P.; Pal, M.; Joshi, R.; Sairam, R.K. Yield, growth and physiological responses of mung bean [*Vigna radiata* (L.) Wilczek] genotypes to waterlogging at vegetative stage. *Physiol. Mol. Biol. Plants* **2013**, *19*, 209–220. [CrossRef]
25. Mustroph, A. Improving Flooding Tolerance of Crop Plants. *Agronomy* **2018**, *8*, 160. [CrossRef]
26. Tomooka, N.; Kaga, A.; Isemura, T.; Vaughan, D. Vigna. In *Wild Crop Relatives: Genomic and Breeding Resources: Legume Crops and Forages*; Kole, C., Ed.; Springer: Berlin/Heidelberg, Germany, 2011; pp. 291–311. [CrossRef]
27. Karuniawan, A.; Iswandi, A.; Kale, P.R.; Heinzemann, J.; Gruneberg, W.J. Vigna vexillata (L.) A. Rich. cultivated as a root crop in Bali and Timor. *Genet. Resour. Crop Evol.* **2006**, *53*, 213–217. [CrossRef]
28. Lawn, R.J.; Watkinson, A.R. Habitats, morphological diversity, and distribution of the genus Vigna Savi in Australia. *Aust. J. Agric. Res.* **2002**, *53*, 1305–1316. [CrossRef]
29. Roecklein, J.C.; Leung, P.S. *A Profile of Economic Plants*; Transaction Publishers: Piscataway/New Brunswick, NJ, USA, 1987.
30. MIller, I.L.; Williams, W.T. Tolerance of some tropical legumes to six months of stimulated waterlogging. *Trop. Glassland* **1981**, *15*, 39–41.
31. Phuphak, S.; Setter, T.L. Adverse effects of waterlogging on growth of lupin and field pea cultivars. In Proceedings of the 5th Australian Agronomy Conference 1989, the University of Western Australia, Perth, Western Australia, 24–29 September 1989.
32. Fukao, T.; Barrera-Figueroa, B.E.; Juntawong, P.; Pena-Castro, J.M. Submergence and Waterlogging Stress in Plants: A Review Highlighting Research Opportunities and Understudied Aspects. *Front. Plant Sci.* **2019**, *10*, 340. [CrossRef]
33. Kortz, A.; Hochholdinger, F.; Yu, P. Cell Type-Specific Transcriptomics of Lateral Root Formation and Plasticity. *Front. Plant Sci.* **2019**, *10*, 21. [CrossRef]
34. Armstrong, W. Aeration in Higher Plants. In *Advances in Botanical Research*; Woolhouse, H.W., Ed.; Academic Press: Cambridge, MA, USA, 1980; Volume 7, pp. 225–332.
35. Mustroph, A.; Zanetti, M.E.; Jang, C.J.; Holtan, H.E.; Repetti, P.P.; Galbraith, D.W.; Girke, T.; Bailey-Serres, J. Profiling translatomes of discrete cell populations resolves altered cellular priorities during hypoxia in Arabidopsis. *Proc. Natl. Acad. Sci. USA* **2009**, *106*, 18843–18848. [CrossRef]
36. Sasidharan, R.; Mustroph, A.; Boonman, A.; Akman, M.; Ammerlaan, A.M.; Breit, T.; Schranz, M.E.; Voesenek, L.A.; van Tienderen, P.H. Root transcript profiling of two Rorippa species reveals gene clusters associated with extreme submergence tolerance. *Plant Physiol.* **2013**, *163*, 1277–1292. [CrossRef]
37. Cai, X.T.; Xu, P.; Zhao, P.X.; Liu, R.; Yu, L.H.; Xiang, C.B. Arabidopsis ERF109 mediates cross-talk between jasmonic acid and auxin biosynthesis during lateral root formation. *Nat. Commun.* **2014**, *5*, 5833. [CrossRef]
38. Negi, S.; Ivanchenko, M.G.; Muday, G.K. Ethylene regulates lateral root formation and auxin transport in Arabidopsis thaliana. *Plant J.* **2008**, *55*, 175–187. [CrossRef]

39. Muday, G.K.; Rahman, A.; Binder, B.M. Auxin and ethylene: Collaborators or competitors? *Trends Plant Sci.* **2012**, *17*, 181–195. [CrossRef]
40. Lewis, D.R.; Negi, S.; Sukumar, P.; Muday, G.K. Ethylene inhibits lateral root development, increases IAA transport and expression of PIN3 and PIN7 auxin efflux carriers. *Development* **2011**, *138*, 3485–3495. [CrossRef]
41. Abbas, M.; Hernandez-Garcia, J.; Pollmann, S.; Samodelov, S.L.; Kolb, M.; Friml, J.; Hammes, U.Z.; Zurbriggen, M.D.; Blazquez, M.A.; Alabadi, D. Auxin methylation is required for differential growth in Arabidopsis. *Proc. Natl. Acad. Sci. USA* **2018**, *115*, 6864–6869. [CrossRef]
42. Spartz, A.K.; Ren, H.; Park, M.Y.; Grandt, K.N.; Lee, S.H.; Murphy, A.S.; Sussman, M.R.; Overvoorde, P.J.; Gray, W.M. SAUR Inhibition of PP2C-D Phosphatases Activates Plasma Membrane H+-ATPases to Promote Cell Expansion in Arabidopsis. *Plant Cell* **2014**, *26*, 2129–2142. [CrossRef]
43. Paponov, I.A.; Paponov, M.; Teale, W.; Menges, M.; Chakrabortee, S.; Murray, J.A.; Palme, K. Comprehensive transcriptome analysis of auxin responses in Arabidopsis. *Mol. Plant* **2008**, *1*, 321–337. [CrossRef]
44. van Mourik, H.; van Dijk, A.D.J.; Stortenbeker, N.; Angenent, G.C.; Bemer, M. Divergent regulation of Arabidopsis SAUR genes: A focus on the SAUR10-clade. *BMC Plant Biol.* **2017**, *17*, 245. [CrossRef]
45. Abel, S.; Nguyen, M.D.; Theologis, A. The PS-IAA4/5-like family of early auxin-inducible mRNAs in Arabidopsis thaliana. *J. Mol. Biol.* **1995**, *251*, 533–549. [CrossRef]
46. Tiwari, S.B.; Wang, X.J.; Hagen, G.; Guilfoyle, T.J. AUX/IAA proteins are active repressors, and their stability and activity are modulated by auxin. *Plant Cell* **2001**, *13*, 2809–2822. [CrossRef]
47. Ulmasov, T.; Murfett, J.; Hagen, G.; Guilfoyle, T.J. Aux/IAA proteins repress expression of reporter genes containing natural and highly active synthetic auxin response elements. *Plant Cell* **1997**, *9*, 1963–1971. [CrossRef]
48. Goh, T.; Joi, S.; Mimura, T.; Fukaki, H. The establishment of asymmetry in Arabidopsis lateral root founder cells is regulated by LBD16/ASL18 and related LBD/ASL proteins. *Development* **2012**, *139*, 883–893. [CrossRef]
49. Cutler, S.; Ghassemian, M.; Bonetta, D.; Cooney, S.; McCourt, P. A protein farnesyl transferase involved in abscisic acid signal transduction in Arabidopsis. *Science* **1996**, *273*, 1239–1241. [CrossRef]
50. Wang, Y.; Ying, J.; Kuzma, M.; Chalifoux, M.; Sample, A.; McArthur, C.; Uchacz, T.; Sarvas, C.; Wan, J.; Dennis, D.T.; et al. Molecular tailoring of farnesylation for plant drought tolerance and yield protection. *Plant J.* **2005**, *43*, 413–424. [CrossRef]
51. Ruppel, N.J.; Kropp, K.N.; Davis, P.A.; Martin, A.E.; Luesse, D.R.; Hangarter, R.P. Mutations in GERANYLGERANYL DIPHOSPHATE SYNTHASE 1 affect chloroplast development in Arabidopsis thaliana (Brassicaceae). *Am. J. Bot.* **2013**, *100*, 2074–2084. [CrossRef]
52. Mitchum, M.G.; Yamaguchi, S.; Hanada, A.; Kuwahara, A.; Yoshioka, Y.; Kato, T.; Tabata, S.; Kamiya, Y.; Sun, T.P. Distinct and overlapping roles of two gibberellin 3-oxidases in Arabidopsis development. *Plant J.* **2006**, *45*, 804–818. [CrossRef]
53. Chaumont, F.; Tyerman, S.D. Aquaporins: Highly regulated channels controlling plant water relations. *Plant Physiol.* **2014**, *164*, 1600–1618. [CrossRef]
54. Hachez, C.; Laloux, T.; Reinhardt, H.; Cavez, D.; Degand, H.; Grefen, C.; De Rycke, R.; Inze, D.; Blatt, M.R.; Russinova, E.; et al. Arabidopsis SNAREs SYP61 and SYP121 coordinate the trafficking of plasma membrane aquaporin PIP2;7 to modulate the cell membrane water permeability. *Plant Cell* **2014**, *26*, 3132–3147. [CrossRef]
55. Ridley, B.L.; O'Neill, M.A.; Mohnen, D.A. Pectins: Structure, biosynthesis, and oligogalacturonide-related signaling. *Phytochemistry* **2001**, *57*, 929–967. [CrossRef]
56. Sobry, S.; Havelange, A.; Van Cutsem, P. Immunocytochemistry of pectins in shoot apical meristems: Consequences for intercellular adhesion. *Protoplasma* **2005**, *225*, 15–22. [CrossRef]
57. Wen, F.; Zhu, Y.; Hawes, M.C. Effect of pectin methylesterase gene expression on pea root development. *Plant Cell* **1999**, *11*, 1129–1140. [CrossRef]
58. Jin, J.; Hewezi, T.; Baum, T.J. Arabidopsis peroxidase AtPRX53 influences cell elongation and susceptibility to Heterodera schachtii. *Plant Signal. Behav.* **2011**, *6*, 1778–1786. [CrossRef]
59. Johansen, D.A. *Plant Microtechnique*; McGraw-Hill Book, Co.: New York, NY, USA, 1940.
60. Bolger, A.M.; Lohse, M.; Usadel, B. Trimmomatic: A flexible trimmer for Illumina sequence data. *Bioinformatics* **2014**, *30*, 2114–2120. [CrossRef]

61. Grabherr, M.G.; Haas, B.J.; Yassour, M.; Levin, J.Z.; Thompson, D.A.; Amit, I.; Adiconis, X.; Fan, L.; Raychowdhury, R.; Zeng, Q.; et al. Full-length transcriptome assembly from RNA-Seq data without a reference genome. *Nat. Biotechnol.* **2011**, *29*, 644–652. [CrossRef]
62. Sirikhachornkit, A.; Suttangkakul, A.; Vuttipongchaikij, S.; Juntawong, P. De novo transcriptome analysis and gene expression profiling of an oleaginous microalga Scenedesmus acutus TISTR8540 during nitrogen deprivation-induced lipid accumulation. *Sci. Rep.* **2018**, *8*, 3668. [CrossRef]
63. Horan, K.; Jang, C.; Bailey-Serres, J.; Mittler, R.; Shelton, C.; Harper, J.F.; Zhu, J.K.; Cushman, J.C.; Gollery, M.; Girke, T. Annotating genes of known and unknown function by large-scale coexpression analysis. *Plant Physiol.* **2008**, *147*, 41–57. [CrossRef]
64. Usadel, B.; Poree, F.; Nagel, A.; Lohse, M.; Czedik-Eysenberg, A.; Stitt, M. A guide to using MapMan to visualize and compare Omics data in plants: A case study in the crop species, Maize. *Plant Cell Environ.* **2009**, *32*, 1211–1229. [CrossRef]
65. Xu, L.; Dong, Z.; Fang, L.; Luo, Y.; Wei, Z.; Guo, H.; Zhang, G.; Gu, Y.Q.; Coleman-Derr, D.; Xia, Q.; et al. OrthoVenn2: A web server for whole-genome comparison and annotation of orthologous clusters across multiple species. *Nucleic Acids Res.* **2019**, *47*, W52–W58. [CrossRef]
66. Liu, Y.; Wei, G.; Xia, Y.; Liu, X.; Tang, J.; Lu, Y.; Lan, H.; Zhang, S.; Li, C.; Cao, M. Comparative transcriptome analysis reveals that tricarboxylic acid cycle-related genes are associated with maize CMS-C fertility restoration. *BMC Plant Biol.* **2018**, *18*, 190. [CrossRef]
67. Zhao, H.; Basu, U.; Kebede, B.; Qu, C.; Li, J.; Rahman, H. Fine mapping of the major QTL for seed coat color in Brassica rapa var. Yellow Sarson by use of NIL populations and transcriptome sequencing for identification of the candidate genes. *PLoS ONE* **2019**, *14*, e0209982. [CrossRef]

© 2019 by the authors. Licensee MDPI, Basel, Switzerland. This article is an open access article distributed under the terms and conditions of the Creative Commons Attribution (CC BY) license (http://creativecommons.org/licenses/by/4.0/).

Article

Transient Heat Waves May Affect the Photosynthetic Capacity of Susceptible Wheat Genotypes Due to Insufficient Photosystem I Photoprotection

Erik Chovancek [1], Marek Zivcak [1,*], Lenka Botyanszka [1], Pavol Hauptvogel [2], Xinghong Yang [3], Svetlana Misheva [4], Sajad Hussain [5] and Marian Brestic [1,6]

1. Department of Plant Physiology, Faculty of Agrobiology and Food Resources, Slovak University of Agriculture, Trieda A. Hlinku 2, 949 76 Nitra, Slovakia
2. National Agricultural and Food Centre, Research Institute of Plant Production, Bratislavska cesta 122, 921 68 Piešt'any, Slovakia
3. College of Life Science, State Key Laboratory of Crop Biology, Shandong Key Laboratory of Crop Biology, Shandong Agricultural University, Taian 271018, China
4. Institute of Plant Physiology and Genetics, Bulgarian Academy of Sciences, 1113 Sofia, Bulgaria
5. Key Laboratory of Crop Ecophysiology and Farming System in Southwest, Ministry of Agriculture, Sichuan Agricultural University, Chengdu 611130, China
6. Department of Botany and Plant Physiology, Faculty of Agrobiology, Food and Natural Resources, Czech University of Life Sciences, 16500 Prague, Czech Republic
* Correspondence: marek.zivcak@uniag.sk; Tel.: +421-37-6414-821

Received: 30 June 2019; Accepted: 5 August 2019; Published: 12 August 2019

Abstract: We assessed the photosynthetic responses of eight wheat varieties in conditions of a simulated heat wave in a transparent plastic tunnel for one week. We found that high temperatures (up to 38 °C at midday and above 20 °C at night) had a negative effect on the photosynthetic functions of the plants and provided differentiation of genotypes through sensitivity to heat. Measurements of gas exchange showed that the simulated heat wave led to a 40% decrease in photosynthetic activity on average in comparison to the control, with an unequal recovery of individual genotypes after a release from stress. Our results indicate that the ability to recover after heat stress was associated with an efficient regulation of linear electron transport and the prevention of over-reduction in the acceptor side of photosystem I.

Keywords: high temperatures; heat stress; photosynthesis; photosystem I; photoprotection; photoinhibition; wheat

1. Introduction

Climate change will bring about an increase in the frequency and intensity of weather extremes, such as heat waves and severe droughts [1,2]. Heat waves (high temperatures for a short time) can significantly reduce the production of grains [3]. Wheat (*Triticum aestivum* L.) is a major staple grain, with a global production of 772 million tons in 2017 [4]. To sustain or even increase production in the future for the rising needs of an increasing human population [5], ongoing adaptation in the form of breeding and suitable agronomic strategies is needed [4].

The optimum growing temperature for wheat is between 17 and 23 °C [6]. A plant is under heat stress when it is exposed to temperatures above an upper threshold for long enough to cause irreversible damage [7]. For wheat, threshold temperatures impacting growth and yield are between 31 and 35 °C [8–10], although some studies have reported high-temperature impacts above even 26 °C [11].

High temperatures cause protein denaturation and aggregation and increase the fluidity of membrane lipids. Indirect heat injuries comprise protein degradation, the inactivation of enzymes in the chloroplast and mitochondria, the inhibition of protein synthesis, and a loss of membrane integrity [12]. These injuries lead to the production of toxic compounds and reactive oxygen species (ROS), reduced ion flux, starvation, and the inhibition of growth [7]. Very high temperatures may cause cell death due to the collapse of cellular organization [13].

Increased temperatures typically lead to a reduction in stomatal conductance (g_s) and thus closure of the stomata [14–16]. However, at high temperatures, g_s might actually increase to avoid a lack of cooling and to avoid dangerously high leaf temperatures [17–19]. Stomatal conductance and net photosynthesis are inhibited by moderate heat stress in many plant species due to decreases in the activation state of rubisco [20,21].

The sites of photochemical reactions are among the first ones to be injured at high temperatures [22]. High temperatures can damage photosystem II (PSII), the oxygen-evolving complex (OEC), and electron transport at both the donor and acceptor sides of PSII in the photosynthetic apparatus [23–26]. PSII is not very stable at high temperatures, and its activity is reduced [27]. Heat stress may cause a dissociation of the OEC and thus an imbalance between the electron flow from the OEC toward the acceptor side of PSII [28].

Not all genotypes within a species have the same ability to cope with heat stress. There is a great deal of variation between and within species, providing opportunities to improve crop heat stress tolerance through genetic means [7]. However, to achieve this goal, contributions from plant physiologists, molecular biologists, and crop breeders are needed. The aim of this study was to provide physiological insights into the effects of a temporary heat wave on photosynthetic functions of wheat leaves, including recovery after heat stress. We focused on the diversity of responses in a group of diverse wheat genotypes of different origins in order to distinguish the photosynthetic responses associated with heat tolerance.

2. Materials and Methods

2.1. Cultivation of Plants

Eight cultivars of winter wheat (*Triticum aestivum* L.) (Equinox (origin: GBR), Thesee (FRA), 16/26 (SVK), GRC 867 (GRC), Roter Samtiger Kolb. (DEU), Unmedpur Mummy (EGY), Dušan (SRB), and AZESVK2009-90 (GEO)) were sown in the middle of November, cultivated at moderate temperature (10–15 °C) for approximately 1 month, and then vernalized in a growth chamber at 5 °C for a photoperiod of 12/12 h (light/dark) for four months, which is the typical duration of the winter period in Slovak wheat production areas. The plants were transplanted during the spring period (May) into pots with standard peat substrate and 5 g of Osmocote fertilizer. The plants were grown individually (one plant per pot) outdoors and were exposed to direct sunlight and natural climatic conditions. The pots were organized in a block with extra border plants, eliminating the effect of borders. The pots were irrigated regularly to prevent dehydration. The high-temperature treatment was started when all plants had fully developed spikes and flag leaves.

2.2. Heat Wave Simulation and Measuring Protocol

The study was carried out at the Slovak University of Agriculture in Nitra, Slovakia. The heat wave was simulated by keeping the plants enclosed under a transparent polyethylene foil tunnel with a high light transmission (>90% of transmitted light at midday) starting in mid-June. Temperatures up to 38 °C were reached inside the tunnel, whereas outside temperatures were between 25 and 30 °C. The measurements were taken between 15 June and 27 June 2018; heat stress was measured on 18, 20, 21, and 22 June (T1 and T2 phase); controls (C) were measured on 19 and 26 June, and the recovery phase (R) was measured on 25 and 27 June. In the recovery phase, the heat-stressed plants were put in control conditions. The measurements of gas exchange and simultaneous measurements of photosystem I (PSI) and photosystem II (PSII) were taken in laboratory conditions.

2.3. Simultaneous Measurements of Gas Exchange and Chlorophyll Fluorescence

The measurements were carried out using an Li-6400 gasometer (LiCor, Lincoln, NE, USA) with simultaneous measurement of chlorophyll fluorescence [29]. The F_0 and F_m values were determined after 15 min of dark adaptation in a measuring head. Then, the sample was exposed to actinic light (1500 µmol photons m^{-2} s^{-1}) at a leaf temperature of 25 °C with a reference CO_2 content of 400 ppm and ambient air humidity. Every 2 min, the gas exchange rate was measured, followed by a saturation pulse and a far-red pulse, for F_0' determination. Then, a CO_2 response curve was applied, starting with a record at 400 ppm and continuing with a stepwise change of levels of CO_2: 300, 250, 200, 150, 100, 50, 400, 600, 800, 1000, 1200, and 1500. The values of gas exchange parameters (CO_2 assimilation rate, A; stomatal conductance, g_s; internal CO_2 concentration, Ci) were calculated directly with a software gas analyzer. Calculations of the chlorophyll fluorescence (ChlF) parameters are described below. Further analyses of the A/Ci curves were performed using the Farquhar–von Caemerer–Berry model [30] edited by Ethier and Livingston [31].

2.4. Simultaneous Measurements of P700 Redox State and Chlorophyll Fluorescence

The state of PSI and PSII photochemistry was measured with a Dual PAM-100 (Walz, Effeltrich, Germany) with a ChlF unit and a P700 dual wavelength (830/875 nm) unit, as described by Klughammer and Schreiber [32]. Saturation pulses (10,000 µmol photons m^{-2} s^{-1}), intended primarily for the determination of ChlF parameters, were also used for the assessment of P700 parameters. Prior to the measurements, the analyzed plants were dark-adapted. After determination of F_0, F_m, and P_m, a moderate light intensity of 134 µmol photons m^{-2} s^{-1} was used to start up the photosynthetic process. After a steady state was reached, a rapid light curve was triggered (with light intensities of 14, 21, 30, 45, 61, 78, 103, 134, 174, 224, 281, 347, 438, 539, 668, 833, 1036, 1295, 1602, and 1960 µmol photons m^{-2} s^{-1} for 30 s at each light intensity). There was a saturation pulse and a far-red pulse for F'_0 determination after 30 s at each light intensity. For the calculation of the ChlF parameters, the following basic values were used: F, F', fluorescence emission from dark-or light-adapted leaf, respectively; F_0, minimum fluorescence from dark-adapted leaf (PSII centers open); F_m, F'_m, maximum fluorescence from dark- or light-adapted leaf, respectively (PSII centers closed); F'_0, minimum fluorescence from light-adapted leaf. The ChlF parameters were calculated as follows [33]: the maximum quantum yield of PSII photochemistry, $F_v/F_m = (F_m - F_0)/F_m$; the actual quantum yield (efficiency) of PSII photochemistry, $\Phi_{PSII} = (F'_m - F')/F'_m$; nonphotochemical quenching, $NPQ = (F_m - F'_m)/F'_m$; the quantum efficiency of nonregulated energy dissipation in PSII, $\Phi_{NO} = 1/(NPQ + 1 + qL (F_m/F_0 - 1))$; the quantum yield of pH-dependent energy dissipation in PSII, $\Phi_{NPQ} = 1 - \Phi_{PSII} - \Phi_{NO}$; and the redox poise of the primary electron acceptor of PSII, Q_A^-/Q_A total $= 1 - qP$. The apparent electron transport rate of PSII photochemistry was calculated by assuming a leaf absorption of 0.84 and a PSII/PSI ratio of 1:1, $ETR_{PSII} = \Phi_{PSII} \times PAR \times 0.84 \times 0.5$.

For the calculation of the P700 parameters, the following basic values were used: P, P700 absorbance at a given light intensity; and P_m, P'_m, the maximum P700 signal measured using a saturation light pulse following short far-red pre-illumination in a dark- or light-adapted state. The P700 parameters were calculated as follows [32]: the effective quantum yield (efficiency) of PSI photochemistry at a given PAR, $\Phi_{PSI} = (P'_m - P)/P_m$; the oxidation status of the PSI donor side, i.e., the fraction of P700 oxidized in a given state, P700$^+$/P700 total $= \Phi_{ND} = P/P_m$; and the reduction status of the PSI acceptor side, i.e., the fraction of overall P700 oxidized in a given state by a saturation pulse due to a lack of electron acceptors, $\Phi_{NA} = (P_m - P'_m)/P_m$. The apparent electron transport rate of the PSI photochemistry was calculated by assuming a leaf absorption of 0.84 and a PSII/PSI ratio of 1:1, $ETR_{PSI} = \Phi_{PSI} \times PAR \times 0.84 \times 0.5$.

2.5. Data Processing and Analysis

The statistical significance of differences was assessed using ANOVA, and post hoc comparisons were performed using Duncan's multiple test (STATISTICA 10, StatSoft, Tulsa, OK, USA). The mean values ± standard error (SE) are presented. At least four plants of each of eight cultivars were measured through gas exchange and chlorophyll fluorescence measurements at four stages. The results of the statistical analyses are not indicated in the graphs, but all of the interpretations are based on these results.

3. Results

The simulation of a heat wave by keeping the plants enclosed under a transparent foil tunnel was effective. The high temperatures (daily maximums at 38 °C and night minimums above 20 °C) had a significant negative effect on the photosynthetic functions of the plants and provided differentiation of genotypes by sensitivity to heat. The gas exchange measurements showed that the simulated heat wave led to a decrease in photosynthetic activity and stomatal conductance of 40% on average in comparison to the control, with a moderate recovery after the relief of stress (Figure 1). Plants that returned to normal conditions (R) after thermal stress showed persistent reductions in photosynthesis due to the several-day high-temperature periods.

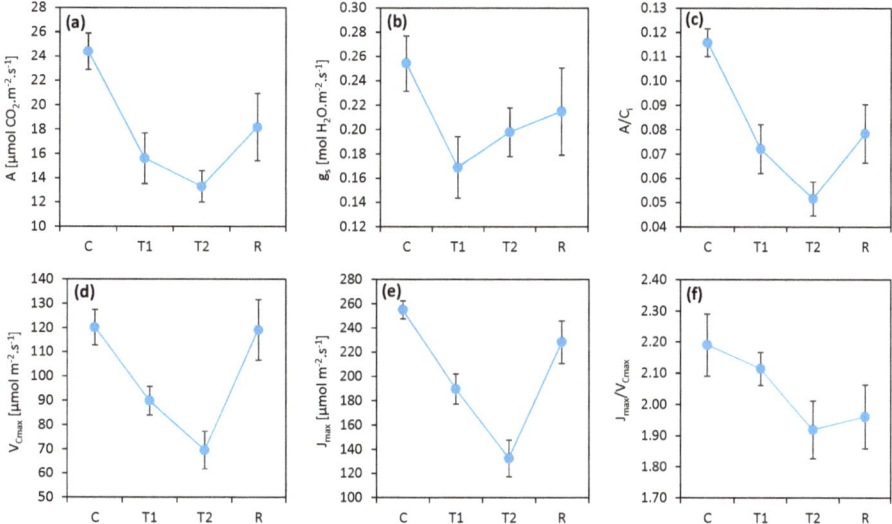

Figure 1. Heat effects on the parameters derived from the gas exchange measurements (**a**) A: photosynthesis rate; (**b**) g_s: stomatal conductance; (**c**) C_i: internal CO_2 concentration; (**d**) V_{Cmax}: maximum carboxylation rate; (**e**) J_{max}: maximum electron transport velocity; (**f**) J_{max}/V_{Cmax} ratio. C: control; T1: thermal effect in the first phase; T2: thermal effect in the second phase; R: recovery phase. The points represent the mean values for all measured wheat plants of all genotypes. The error bars represent the standard error of the means.

The ratio between the CO_2 assimilation rate and the CO_2 concentration in the intercellular spaces in the leaf (A/C_i) expresses the efficiency of CO_2 utilization by the photosynthetic apparatus and served to verify whether the monitored decrease in photosynthesis was caused by the closing of stomata or nonstomatal causes, mostly decreases in photosynthetic enzyme activities [29]. The decrease in A/C_i followed the trend of decreasing photosynthesis, which suggests that the closure of the stomata had only a marginal effect and that the nonstomatal limitation of photosynthesis was dominant. In addition, the trend of decreasing CO_2 assimilation during the heat stress period (from T1 to T2) was opposite to that seen in stomatal conductance, suggesting a minor role of stomatal closure in the decrease in CO_2 assimilation caused by the heat wave.

Further analysis of the A/C$_i$ curves using the Farquhar–von Caemerer–Berry model [30] edited by Ethier and Livingston [31] helped reveal the partial limitations of the assimilation process. V_{Cmax}, or maximum carboxylation rate, represented a limitation in rubisco enzyme activity, while J_{max}, or maximum electron transport velocity, represented a limitation of the primary photosynthesis product in C3 plants, RuBP [34]. The results of these parameters corresponded quite well with photosynthesis, which means that they participated at the same proportion in the limitation of photosynthesis under the influence of heat. Thus, the photosynthetic limitation identified by the A/C$_i$ parameter was due both to a decrease in rubisco total activity and to a decrease in RuBP regeneration, which is usually attributed to limited electron transport in chloroplasts [30]. It is worth mentioning that the trend of V_{Cmax} was not the same as J_{max}. The maximum electron transport (J_{max}) was more affected by heat stress compared to the carboxylation activity of rubisco (V_{Cmax}) in conditions of a heat wave. Moreover, whereas V_{Cmax} almost completely recovered, J_{max} was still decreased after the heat wave. This trend was well illustrated by the J_{max} to V_{Cmax} ratio, which decreased due to heat stress, and no recovery was observed after heat stress ended.

The high temperatures inflicted a nonstomatal limitation of photosynthesis. This effect was proven by the decrease in rubisco activity as well as by the parameters of photochemistry.

By comparing different varieties (Figure 2), we found that there was considerable variability in the response to heat and in the ability to recover after stress. The highest photosynthesis levels were in the varieties Dušan and Roter Samtiger Kolbenweizen before the heat wave. At the same time, these were varieties that showed the highest rate of photosynthesis decrease after two (T1) and four (T2) days of temperature stress. These two varieties differed significantly in their ability to regenerate the photosynthetic apparatus. While the leaves of Roter Samtiger Kolbenweizen died rapidly after thermal stress (they did not regenerate), Dušan regenerated very well, similarly to genotypes GRC 867, AZESVK2009-90, and Unmedpur Mummy. The least influence of heat in phases T1 and T2 was observed in the variety GRC 867. The heat-stressed plants of genotype Thesee showed poor recovery and started premature senescence of the leaves, which did not occur in the control plants. Similar, but less evident, trends were also observed in cv. Equinox and 16/26. The values of stomatal conductance (g_s) decreased, similarly to CO_2 assimilation. However, the decrease in the A/Ci ratio indicated that stomata closure was not the major reason for the photosynthetic decline in stress and recovery conditions. Values for the maximum rate of carboxylation derived from the initial slope of the A/Ci curve (V_{Cmax}) showed a very similar trend to the observed CO_2 assimilation rate.

The measurements of photosynthetic quantum yields of both photosystems showed a decrease in the activity of both photosystems in reaction to high temperatures. The high temperatures decreased the activity of both photosystems during stress as well as after stress. The heat simulated in the tunnel led to a decrease in the quantum efficiency of both PSII and PSI (Figure 3) by approximately 40–50%. The quantum yield of regulated nonphotochemical quenching (Φ_{NPQ}) was lower in heat-stressed plants and, interestingly, higher in recovered plants than in the control. As a result of the decrease in Φ_{PSII} and Φ_{NPQ}, we observed very high values in the fraction of nonregulated (passive) nonphotochemical dissipation (Φ_{NO}). The decrease in Φ_{PSI} was caused by an increase in the acceptor side limitation (Φ_{NA}). Interestingly, the values of the quantum yield of nonphotochemical quenching of PSI caused by the donor side limitation (Φ_{ND}) did not change significantly.

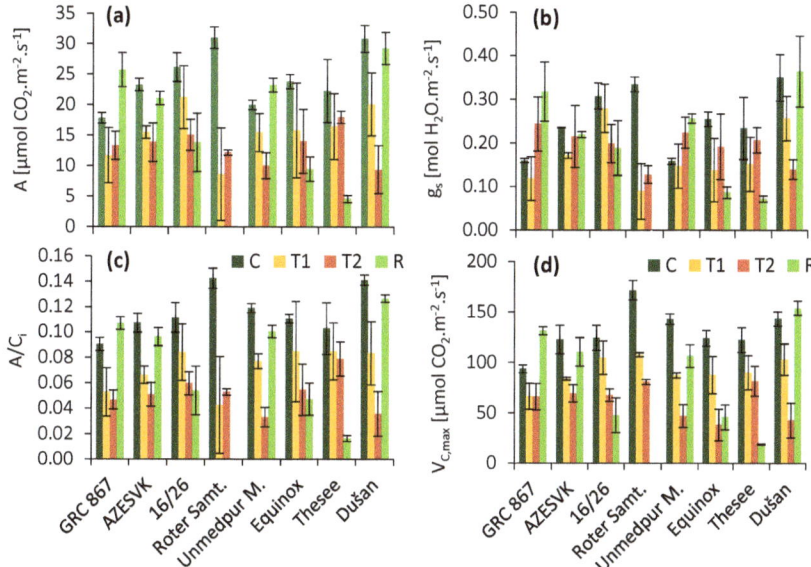

Figure 2. Heat effect on parameters measured by Li-6400 (**a**) A, photosynthesis rate; (**b**) g_s, stomatal conductance; (**c**) A/C_i, photosynthetic rate per unit of internal CO_2 concentration; (**d**) V_{Cmax}, maximum rate of carboxylation based on analyses of A/Ci curves; C, control; T1, thermal effect in the first phase; T2, thermal effect in the second phase; R, recovery phase. Mean values ± SE are presented.

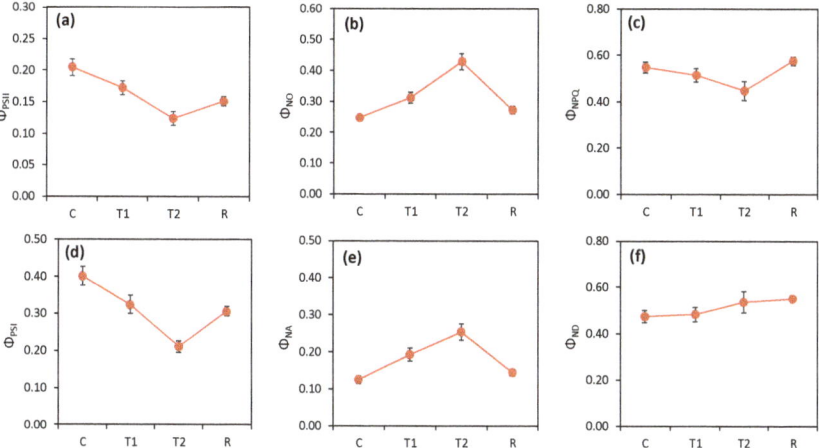

Figure 3. Heat effect on parameters measured by Dual-PAM (the average from all varieties): C, control; T1, thermal effect in the first phase; T2, thermal effect in the second phase; R, recovery phase. (**a**) The effective quantum yield of photosystem II (PSII) (Φ_{PSII}); (**b**) the fraction of energy captured by PSII passively dissipated in the form of heat and fluorescence (Φ_{NO}); (**c**) the quantum yield of regulated nonphotochemical quenching in PSII (Φ_{NPQ}); (**d**) the effective quantum yield of PSI (Φ_{PSI}); (**e**) the quantum yield of PSI nonphotochemical quenching caused by the acceptor side limitation, i.e., the fraction of overall P700 that could not be oxidized in a given state (Φ_{NA}); (**f**) the quantum yield of PSI nonphotochemical quenching caused by the donor side limitation, i.e., the fraction of overall P700 that was oxidized in a given state (Φ_{ND}). The points represent the mean values for all measured wheat plants of all genotypes. The error bars represent the standard error of the means.

The trends in the photosynthetic electron transport rate (ETR, Figure 4a) as well as the value of photosynthetic assimilation in different wheat genotypes (Figure 2) clearly indicated differences in the ability to recover after the heat stress period ended. The leaves of the genotype Roter Samtiger, especially, became necrotic and dried as a consequence of heat stress. Moreover, some other genotypes expressed insufficient recovery. We also analyzed how the genotypes were able to downregulate over-reduction in the PSI acceptor side (Figure 4b). We observed that the same genotypes were characterized by an over-reduction in the PSI acceptor side during the heat stress period, whereas this was not so obvious in well-recovering genotypes. These results were also evident in the values of fluorescence and the PSI parameters measured in a dark-adapted state. It was evident that the decrease in F_v/F_m was most severe in the Roter Samtiger cultivar and that the values were also not recovered in the Thesee cultivar. Different trends were observed in parameter P_m, representing a maximum amplitude of P700 kinetics, in which we found only partial or no recovery (cv. Rotter Samtiger, Thesee) after relief from heat stress. Moreover, the genotypes differed in the severity of P_m decrease and the level of P_m recovery.

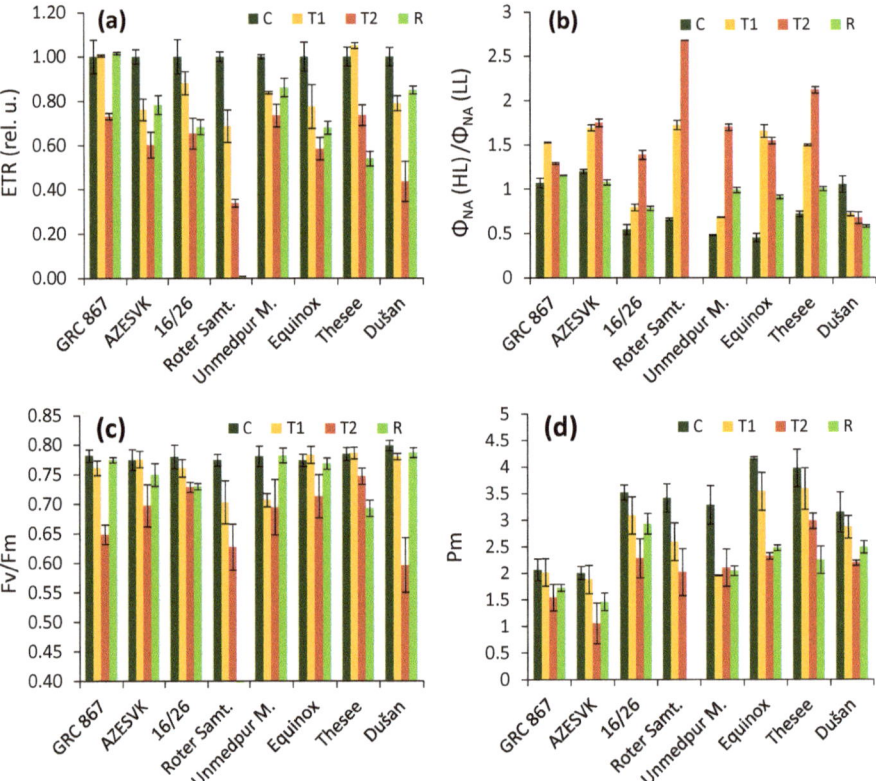

Figure 4. Effects of heat stress on parameters measured by Dual-PAM in different varieties: C, control; T1, thermal effect in the first phase; T2, thermal effect in the second phase; R, recovery phase. (**a**) The relative values of the electron transport rate (the average value of the control plants of each genotype equals 1). (**b**) The ratio of the acceptor side limitation measured in high light (HL, ~2000 µmol m^{-2} s^{-1}) and the value measured in low light (LL, ~40 µmol m^{-2} s^{-1}). (**c**) Maximum quantum efficiency of PSII photochemistry. (**d**) The maximum amplitude of P700 kinetics. Average values ± SE are presented.

Detailed analyses of the light response of the PSI acceptor side state (Figure 5) indicated that the well-regenerating genotype GRC 867 was able to efficiently downregulate electron transport, keeping the PSI acceptor side reduction low, even in very high light. The photoprotective capacity of genotypes became insufficient at light intensities over ~600 µmol m^{-2} s^{-1}. The regulation of the electron transport in the genotype Roter Samtiger failed at a PAR below 300 µmol m^{-2} s^{-1}.

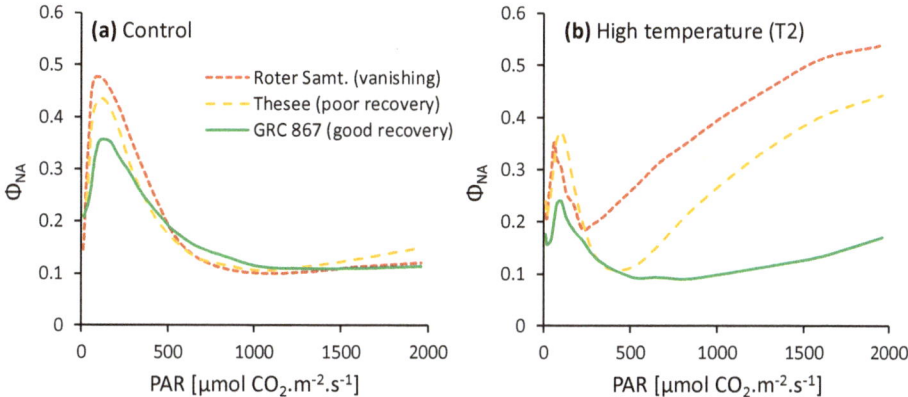

Figure 5. Examples of the light response curves of the PSI acceptor side limitation parameter (Φ_{NA}) measured in control plants (**a**) and in plants exposed to the heat wave (**b**) of three genotypes differing in their capacity to recover after the withdrawal of heat stress.

4. Discussion

There are several target sites for elevated temperature-induced damage, such as the CO_2 fixation system, photophosphorylation, the electron transport chain, and the OEC [35]. The enzymes of the Calvin–Benson cycle are heat-labile. This means that the carbon assimilation system is sensitive to elevated temperatures and is strongly inhibited at moderate thermal stress [36]. These inhibiting effects are mostly observed when measured directly at high-temperature conditions [35,36], which was not the case in our experiments, in which photosynthesis was measured at a normal temperature (25 °C) at least 12 h after the last exposure to high temperature. The heat effects observed in our study were consequences and not the instantaneous effects of heat stress on photosynthesis. In this respect, it is interesting that the post-stress effects observed in our study were similar to the instantaneous effects well known from other studies. One of them was the decrease in rubisco activity indicated by the decrease in V_{Cmax}, which was associated with a decrease in RuBP regeneration and the limitation in photosynthesis represented by the parameter J_{max} (Figure 1). As direct damage to rubisco was not likely in the conditions of our experiments, an inhibition of the enzyme (especially rubisco activase) caused by its sensitivity to moderately high temperatures [37,38] is more probable. There is a hypothesis that a decrease in rubisco activation represents a protective mechanism against a critical decrease in the transthylakoid proton gradient in high-temperature conditions to prevent the collapse of photoprotective functions, with fatal consequences associated mainly with an uncontrolled increase of oxidative stress [36,39]. If long-term effects from high temperatures on photoprotection occur, the downregulation of enzyme activities might be needed even after temperatures return to normal. Moreover, the ratio of J_{max} to V_{Cmax} decreased due to stress (Figure 1f), which may indicate that electron transport-related processes were affected more than the carboxylation activity of rubisco was. Therefore, we focused on the processes associated with photosynthetic electron transport.

An analysis of basic chlorophyll fluorescence and P700 parameters in nonstressed plants (Figure 3) confirmed this expectation and identified the sustaining effects of high temperatures on PSII and PSI photochemistry, including the photoinhibition of PSI, which were similar to the effects observed in our previous study on wheat exposed to high temperature [40]. We previously showed that PSI photoinhibition had major effects on carbon assimilation [41] and photoprotection [42]. The trends in the parameters measured by simultaneous chlorophyll fluorescence and P700 (Figure 3) indirectly (e.g., a decrease in Φ_{NPQ} despite a decrease in Φ_{PSII}) or directly (increase in Φ_{NA}) pointed to damage to PSI functions. Thus, our results support the hypothesis that the decrease in photosynthetic assimilation was associated with a decrease in photochemical activities.

In addition to the general trends, we observed some variance among the observed genotypes. The most important were the differences in recovery several days after heat stress relief, which was evident in the gas exchange as well as in the photochemical responses (Figures 2 and 4). We observed an extreme response in genotype Roter Samtiger, in which heat stress led to severe necrosis and the death of leaves. In addition, we observed low recovery in some other genotypes, especially in the genotype Thesee. Interestingly, in these genotypes, we observed a very high level of over-reduction in the PSI acceptor side in high light conditions (high values of parameter Φ_{NA}), whereas in well-regenerating genotypes, the parameter Φ_{NA} was kept low in high light. It has previously been shown that a high Φ_{NA} is an indicator of over-reduction in the PSI acceptor side [43–46], which leads to the excessive production of ROS in PSI [47–49]. Such a situation is known to be responsible for PSI photoinhibition in vivo [50,51]. PSI photoinhibition is characterized by very low recovery, and in some cases, PSI damage is not completely reversible [52,53]. Most photoinhibited PSI reaction center complexes are not repaired, but degrade after photoinhibition together with their binding chlorophylls [54]. This is a completely different situation compared to PSII, which is able to quickly recover. In the most sensitive genotypes, we observed a loss of ability to downregulate linear electron transport even at moderate light intensities, which might have easily resulted in leaf damage due to the accumulation of ROS in tissues. This explains the necrosis of leaves, which led to their premature death. A high ROS production could trigger the processes of early senescence associated with a decrease in the photosynthetic capacity of the leaves in sensitive genotypes. In the genotypes that had well-regulated electron transport, early senescence was not observed.

Considering the possible practical relevance of the Φ_{NA} parameter measure during heat stress as an indicator of heat-sensitive genotypes, it must be noted that only the values of the Φ_{NA} or Φ_{NA} (HL) to Φ_{NA} (LL) ratio (Figure 4) could not fully explain the level of recovery of photosynthetic capacity after the heat wave in all genotypes. For example, the values of the Φ_{NA} (HL) to Φ_{NA} (LL) ratio in the heat stress stage in cv. Equinox and AZESVK were similar, but AZESVK recovered better. In this respect, the changes in values in the Φ_{NA} (HL) to Φ_{NA} (LL) ratio between control conditions and heat wave conditions seemed to be more indicative. It is obvious that the sensitive cultivars showed a higher change in the values of the Φ_{NA} (HL) to Φ_{NA} (LL) ratio compared to the more resistant cultivars.

Rubisco activation, assessed by the initial slopes of the A/Ci curves [55], has previously been found to be a possible reason for the improper regulation of electron transport, identified at the level of PSI. When rubisco activity decreased, Φ_{NA} increased, and Φ_{ND} was suppressed [56]. The general trend of Φ_{NA} increase (Figures 4 and 5) could have been caused primarily by the difference in rubisco activation and a decrease in the need for an electron transport in photosynthesis. On the other hand, the differences in carboxylation activity observed in our study cannot explain the different trends of Φ_{NA} shown in Figure 5. Therefore, we suggest that the different susceptibilities to high temperatures observed in our study were not directly associated with the rubisco activation state.

One interesting trend observed in our study was a higher decrease in the capacity of the electron transport rate (represented by J_{max}) compared to the carboxylation capacity (V_{Cmax}), both in heat stress and recovery periods. The values of P_m corresponded better to the records of the CO_2 assimilation rate or the ETR compared to F_v/F_m (Figure 4), especially during the recovery period. It is obvious that whereas PSII recovered very well (in six of eight genotypes), the recovery of PSI was much lower and was insufficient in most of the plants. Thus, the lower activity of PSI could have been responsible (at least partially) for the decrease in the electron transport capacity after the transient heat wave period.

Overall, the results suggest that the proper regulation of electron transport and the efficient photoprotection of PSI against photoinhibition were crucial in preventing negative post-stress effects after plants were exposed to short transient periods of high temperatures, which commonly occurs during the crop vegetative period. This may be important for breeding strategies, as the probability of heat waves will increase due to climate change.

Author Contributions: Conceptualization, M.B., X.Y., and M.Z.; data curation, M.Z.; formal analysis, M.Z. and E.C.; funding acquisition, M.B. and M.Z.; investigation, E.C., L.B., and M.Z.; methodology, M.Z.; project administration, M.B.; resources, P.H.; supervision, M.Z.; validation, X.Y., P.H., and S.P.M.; visualization, E.C. and M.Z.; writing—original draft, E.C.; writing—review and editing, M.Z., M.B., X.Y., P.H., S.H., and S.P.M.

Funding: This research was funded by the grants APVV-15-0721 and VEGA-1-0831-17.

Conflicts of Interest: The authors declare no conflicts of interest.

References

1. Pachauri, R.K.; Meyer, L.A. (Eds.) Intergovernmental Panel on Climate Change. In *Climate Change 2014: Synthesis Report*; IPCC: Geneva, Switzerland, 2014.
2. Duan, H.; Wu, J.; Huang, G.; Zhou, S.; Liu, W.; Liao, Y.; Yang, X.; Xiao, Z.; Fan, H. Individual and interactive effects of drought and heat on leaf physiology of seedlings in an economically important crop. *AoB Plants* **2017**, *9*, plw090. [CrossRef] [PubMed]
3. Nuttall, J.G.; Barlow, K.M.; Delahunty, A.J.; Christy, B.P.; O'Leary, G.J. Acute high temperature response in wheat. *Agron. J.* **2018**, *110*, 1296–1308. [CrossRef]
4. FAOSTAT. Production of Wheat in World. Available online: http://www.fao.org/faostat/en/#data/QC/visualize (accessed on 23 February 2019).
5. Slafer, G.A.; Savin, R.; Sadras, V.O. Coarse and fine regulation of wheat yield components in response to genotype and environment. *Field Crops Res.* **2014**, *157*, 71–83. [CrossRef]
6. Shanmugam, S.; Kjaer, K.H.; Ottosen, C.O.; Rosenqvist, E.; Kumari Sharma, D.; Wollenweber, B. The alleviating effect of elevated CO_2 on heat stress susceptibility of two wheat (*Triticum aestivum* L.) cultivars. *J. Agron. Crop. Sci.* **2013**, *199*, 340–350. [CrossRef]
7. Wahid, A.; Gelani, S.; Ashraf, M.; Foolad, M.R. Heat tolerance in plants: An overview. *Environ. Exp. Bot.* **2007**, *61*, 199–223. [CrossRef]
8. Ferris, R.; Ellis, R.H.; Wheeler, T.R.; Hadley, P. Effect of high temperature stress at anthesis on grain yield and biomass of field-grown crops of wheat. *Ann. Bot.* **1998**, *82*, 631–639. [CrossRef]
9. Barnabas, B.; Jager, K.; Feher, A. The effect of drought and heat stress on reproductive processes in cereals. *Plant Cell Environ.* **2008**, *31*, 11–38. [CrossRef]
10. Fischer, R.A. Wheat physiology: A review of recent developments. *Crop Pasture Sci.* **2011**, *62*, 95–114. [CrossRef]
11. Stone, P.; Nicolas, M. Wheat cultivars vary widely in their responses of grain yield and quality to short periods of post-anthesis heat stress. *Funct. Plant Biol.* **1994**, *21*, 887–900. [CrossRef]
12. Howarth, C.J. Genetic improvements of tolerance to high temperature. In *Abiotic Stresses*; Ashraf, M., Harris, P., Eds.; CRC Press: Boca Raton, FL, USA, 2005; pp. 299–322.
13. Schoeffl, F.; Prandl, R.; Reindl, A. Molecular responses to heat stress. In *Molecular Responses to Cold, Drought, Heat and Salt Stress in Higher Plants*; Shinozaki, K., Yamaguchi-Shinozaki, K., Eds.; R.G.Landes Co.: Austin, TX, USA, 1999; pp. 81–98.
14. Way, D.A.; Oren, R.; Kroner, Y. The space-time continuum: The effects of elevated CO_2 and temperature and the importance of scaling. *Plant Cell Environ.* **2015**, *38*, 991–1007. [CrossRef]
15. Slot, M.; Winter, K. In Situ temperature relationships of biochemical and stomatal controls of photosynthesis in four lowland tropical tree species. *Plant Cell Environ.* **2017**, *40*, 3055–3068. [CrossRef] [PubMed]
16. Fauset, S.; Oliveira, L.; Buckeridge, M.S.; Foyer, C.H.; Galbraith, D.; Tiwari, R.; Gloor, M. Contrasting responses of stomatal conductance and photosynthetic capacity to warming and elevated CO_2 in the tropical tree species *Alchornea glandulosa* under heatwave conditions. *Environ. Exp. Bot.* **2019**, *158*, 28–39. [CrossRef]

17. Drake, J.E.; Tjoelker, M.G.; Vårhammar, A.; Medlyn, B.E.; Reich, P.B.; Leigh, A.; Barton, C.V.M. Trees tolerate an extreme heatwave via sustained transpirational cooling and increased leaf thermal tolerance. *Glob. Chang. Biol.* **2018**, *24*, 2390–2402. [CrossRef] [PubMed]
18. Urban, J.; Ingwers, M.W.; McGuire, M.A.; Teskey, R.O. Increase in leaf temperature opens stomata and decouples net photosynthesis from stomatal conductance in *Pinus taeda* and *Populus deltoides* x *nigra*. *J. Exp. Bot.* **2017**, *68*, 1757–1767. [CrossRef] [PubMed]
19. Slot, M.; Winter, K. Photosynthetic acclimation to warming in tropical forest tree seedlings. *J. Exp. Bot.* **2017**, *68*, 2275–2284. [CrossRef] [PubMed]
20. Crafts-Brander, S.J.; Salvucci, M.E. Sensitivity to photosynthesis in the C4 plant, maize to heat stress. *Plant Cell* **2002**, *12*, 54–68.
21. Morales, D.; Rodríguez, P.; Dell'amico, J.; Nicolás, E.; Torrecillas, A.; Sánchez-Blanco, M.J. High-temperature preconditioning and thermal shock imposition affects water relations, gas exchange and root hydraulic conductivity in tomato. *Biol. Plant.* **2003**, *47*, 203–208. [CrossRef]
22. Wise, R.R.; Olson, A.J.; Schrader, S.M.; Sharkey, T.D. Electron transport is the functional limitaion of photosynthesis in field-grown *Pima* cotton plants at high temperature. *Plant Cell Environ.* **2004**, *27*, 717–724. [CrossRef]
23. Tóth, S.Z.; Schansker, G.; Kissimon, J.; Kovács, L.; Garab, G.; Strasser, R.J. Biophysical studies of photosystem II-related recovery processes after a heat pulse in barley seedling (*Hordeum vulgare*). *J. Plant Physiol.* **2005**, *162*, 181–194. [CrossRef]
24. Lazár, D. The polyphasic chlorophyll a fluorescence rise measured under high intensity of exciting light. *Funct. Plant Biol.* **2006**, *33*, 9–30. [CrossRef]
25. Chen, L.S.; Li, P.; Cheng, L. Effects of high temperature coupled with high light on the balance between photooxidation and photoprotection in the sun-exposed peel of apple. *Planta* **2008**, *228*, 745–756. [CrossRef]
26. Li, P.; Cheng, L.; Gao, H.; Jiang, C.; Peng, T. Heterogenous behavior of PSII in soybean (*Glycine max*) leaves with identical PSII photochemistry efficiency under different high temperature treatments. *J. Plant Physiol.* **2009**, *166*, 1607–1615. [CrossRef]
27. Camejo, D.; Rodríguez, P.; Morales, M.A.; Dellamico, J.M.; Torrecillas, A.; Alarcón, J.J. High temperature effects on photosynthetic activity of two tomato cultivars with different heat susceptibility. *J. Plant Physiol.* **2005**, *162*, 281–289. [CrossRef]
28. De Ronde, J.A.D.; Cress, W.A.; Kruger, G.H.J.; Strasser, R.J.; Staden, J.V. Photosynthetic response of transgenic soybean plants containing an *Arabidopsis* P5CR gene, during heat and drought stress. *J. Plant Physiol.* **2004**, *61*, 1211–1244. [CrossRef]
29. Zivcak, M.; Brestic, M.; Balatova, Z.; Drevenakova, P.; Olsovska, K.; Kalaji, H.M.; Allakhverdiev, S.I. Photosynthetic electron transport and specific photoprotective responses in wheat leaves under drought stress. *Photosynt. Res.* **2013**, *117*, 529–546. [CrossRef]
30. Farquhar, G.D.; von Caemmerer, S.; Berry, J.A. A biochemical model of photosynthetic CO_2 assimilation in leaves of C3 species. *Planta* **1980**, *149*, 78–90. [CrossRef]
31. Ethier, G.J.; Livingston, N.J. On the need to incorporate sensitivity to CO_2 transfer conductance into the Farquhar–von Caemmerer-Berry leaf photosynthesis model. *Plant Cell Environ.* **2004**, *27*, 137–153. [CrossRef]
32. Klughammer, C.; Schreiber, U. Saturation pulse method for assessment of energy conversion in PSI. *Planta* **1994**, *192*, 261–268. [CrossRef]
33. Baker, N.R. Chlorophyll fluorescence: A probe of photosynthesis in vivo. *Ann. Rev. Plant Biol.* **2008**, *59*, 89–113. [CrossRef]
34. Brestic, M.; Zivcak, M.; Hauptvogel, P.; Misheva, S.; Kocheva, K.; Yang, X.; Li, X.; Allakhverdiev, S.I. Wheat plant selection for high yields entailed improvement of leaf anatomical and biochemical traits including tolerance to non-optimal temperature conditions. *Photosynt. Res.* **2018**, *136*, 245–255. [CrossRef]
35. Allakhverdiev, S.I.; Kreslavski, V.D.; Klimov, V.V.; Los, D.A.; Carpentier, R.; Mohanty, P. Heat stress: An overview of molecular responses in photosynthesis. *Photosynt. Res.* **2008**, *98*, 541. [CrossRef]
36. Sharkey, T.D. Effects of moderate heat stress on photosynthesis: Importance of thylakoid reactions, rubisco deactivation, reactive oxygen species, and thermotolerance provided by isoprene. *Plant Cell Environ.* **2005**, *28*, 269–277. [CrossRef]
37. Salvucci, M.E.; Crafts-Brandner, S.J. Inhibition of photosynthesis by heat stress: The activation state of Rubisco as a limiting factor in photosynthesis. *Physiol. Plantarum* **2004**, *120*, 179–186. [CrossRef]

38. Haldimann, P.; Feller, U. Growth at moderately elevated temperature alters the physiological response of the photosynthetic apparatus to heat stress in pea (*Pisum sativum* L.) leaves. *Plant Cell Environ.* **2005**, *28*, 302–317. [CrossRef]
39. Zhang, R.; Sharkey, T.D. Photosynthetic electron transport and proton flux under moderate heat stress. *Photosynth. Res.* **2009**, *100*, 29–43. [CrossRef]
40. Brestic, M.; Zivcak, M.; Kunderlikova, K.; Allakhverdiev, S.I. High temperature specifically affects the photoprotective responses of chlorophyll b-deficient wheat mutant lines. *Photosynth. Res.* **2016**, *130*, 251–266. [CrossRef]
41. Zivcak, M.; Brestic, M.; Kunderlikova, K.; Sytar, O.; Allakhverdiev, S.I. Repetitive light pulse-induced photoinhibition of photosystem I severely affects CO_2 assimilation and photoprotection in wheat leaves. *Photosynth. Res.* **2015**, *126*, 449–463. [CrossRef]
42. Brestic, M.; Zivcak, M.; Kunderlikova, K.; Sytar, O.; Shao, H.; Kalaji, H.M.; Allakhverdiev, S.I. Low PSI content limits the photoprotection of PSI and PSII in early growth stages of chlorophyll b-deficient wheat mutant lines. *Photosynth. Res.* **2015**, *125*, 151–166. [CrossRef]
43. Huang, W.; Yang, Y.J.; Zhang, S.B. Specific roles of cyclic electron flow around photosystem I in photosynthetic regulation in immature and mature leaves. *J. Plant Phys.* **2017**, *209*, 76–83. [CrossRef]
44. Takagi, D.; Miyake, C. Proton gradient regulation 5 supports linear electron flow to oxidize photosystem I. *Physiologia Plantarum* **2018**, *164*, 337–348. [CrossRef]
45. Takagi, D.; Takumi, S.; Miyake, C. Growth light environment changes the sensitivity of photosystem I photoinhibition depending on common wheat cultivars. *Front. Plant Sci.* **2019**, *10*, 686. [CrossRef]
46. Wada, S.; Takagi, D.; Miyake, C.; Makino, A.; Suzuki, Y. Responses of the photosynthetic electron transport reactions stimulate the oxidation of the reaction center chlorophyll of photosystem I, p700, under drought and high temperatures in rice. *Int. J. Mol. Sci.* **2019**, *20*, 2068. [CrossRef]
47. Schmitt, F.J.; Renger, G.; Friedrich, T.; Kreslavski, V.D.; Zharmukhamedov, S.K.; Los, D.A.; Kuznetsov, V.V.; Allakhverdiev, S.I. Reactive oxygen species: Re-evaluation of generation, monitoring and role in stress-signalling in phototrophic organisms. *Biochim. Biophys. Acta* **2014**, *1837*, 835–848. [CrossRef]
48. Takagi, D.; Amako, K.; Hashiguchi, M.; Fukaki, H.; Ishizaki, K.; Goh, T.; Sawa, S. Chloroplastic ATP synthase builds up a proton motive force preventing production of reactive oxygen species in photosystem I. *Plant J.* **2017**, *91*, 306–324. [CrossRef]
49. Huang, W.; Tikkanen, M.; Zhang, S.B. Photoinhibition of photosystem I in *Nephrolepis falciformis* depends on reactive oxygen species generated in the chloroplast stroma. *Photosynth. Res.* **2018**, *137*, 129–140. [CrossRef]
50. Tikkanen, M.; Mekala, N.R.; Aro, E.M. Photosystem II photoinhibition-repair cycle protects Photosystem I from irreversible damage. *Biochim. Biophys. Acta* **2014**, *1837*, 210–215. [CrossRef]
51. Kono, M.; Noguchi, K.; Terashima, I. Roles of the cyclic electron flow around PSI (CEF-PSI) and O_2-dependent alternative pathways in regulation of the photosynthetic electron flow in short-term fluctuating light in *Arabidopsis thaliana*. *Plant Cell Physiol.* **2014**, *55*, 990–1004. [CrossRef]
52. Kudoh, H.; Sonoike, K. Irreversible damage to photosystem I by chilling in the light: Cause of the degradation of chlorophyll after returning to normal growth temperature. *Planta* **2002**, *215*, 541–548. [CrossRef]
53. Zivcak, M.; Brestic, M.; Kunderlikova, K.; Olsovska, K.; Allakhverdiev, S.I. Effect of photosystem I inactivation on chlorophyll a fluorescence induction in wheat leaves: Does activity of photosystem I play any role in OJIP rise? *J. Photochem. Photobiol.* **2015**, *152*, 318–324. [CrossRef]
54. Sonoike, K. Photoinhibition of photosystem I. *Physiol. Plant* **2011**, *142*, 56–64. [CrossRef]
55. Fukayama, H.; Ueguchi, C.; Nishikawa, K.; Katoh, N.; Ishikawa, C.; Masumoto, C.; Hatanaka, T.; Misoo, S. Overexpression of Rubisco activase decreases the photosynthetic CO_2 assimilation rate by reducing Rubisco content in rice leaves. *Plant Cell Physiol.* **2012**, *53*, 976–986. [CrossRef]
56. Wada, S.; Suzuki, Y.; Takagi, D.; Miyake, C.; Makino, A. Effects of genetic manipulation of the activity of photorespiration on the redox state of photosystem I and its robustness against excess light stress under CO_2-limited conditions in rice. *Photosynth. Res.* **2018**, *137*, 431–441. [CrossRef]

© 2019 by the authors. Licensee MDPI, Basel, Switzerland. This article is an open access article distributed under the terms and conditions of the Creative Commons Attribution (CC BY) license (http://creativecommons.org/licenses/by/4.0/).

MDPI
St. Alban-Anlage 66
4052 Basel
Switzerland
Tel. +41 61 683 77 34
Fax +41 61 302 89 18
www.mdpi.com

Plants Editorial Office
E-mail: plants@mdpi.com
www.mdpi.com/journal/plants

www.ingramcontent.com/pod-product-compliance
Lightning Source LLC
LaVergne TN
LVHW071955080526
838202LV00064B/6752